SpringerBriefs in Molecular Science

Biobased Polymers

Series Editor

Patrick Navard, CNRS/Mines ParisTech, Sophia Antipolis, France

Published under the auspices of EPNOE*, *Springerbriefs in Biobased polymers* covers all aspects of biobased polymer science, from the basis of this field starting from the living species in which they are synthetized (such as genetics, agronomy, plant biology) to the many applications they are used in (such as food, feed, engineering, construction, health, ..) through to isolation and characterization, biosynthesis, biodegradation, chemical modifications, physical, chemical, mechanical and structural characterizations or biomimetic applications. All biobased polymers in all application sectors are welcome, either those produced in living species (like polysaccharides, proteins, lignin, ..) or those that are rebuilt by chemists as in the case of many bioplastics.

Under the editorship of Patrick Navard and a panel of experts, the series will include contributions from many of the world's most authoritative biobased polymer scientists and professionals. Readers will gain an understanding of how given biobased polymers are made and what they can be used for. They will also be able to widen their knowledge and find new opportunities due to the multidisciplinary contributions.

This series is aimed at advanced undergraduates, academic and industrial researchers and professionals studying or using biobased polymers. Each brief will bear a general introduction enabling any reader to understand its topic.

EPNOE The European Polysaccharide Network of Excellence (www.epnoe.eu) is a research and education network connecting academic, research institutions and companies focusing on polysaccharides and polysaccharide-related research and business.

More information about this subseries at http://www.springer.com/series/15056

Geeta Gahlawat

Polyhydroxyalkanoates Biopolymers

Production Strategies

 Springer

Geeta Gahlawat
Department of Microbiology
Panjab University
Chandigarh, India

ISSN 2191-5407   ISSN 2191-5415 (electronic)
SpringerBriefs in Molecular Science
ISSN 2510-3407   ISSN 2510-3415 (electronic)
Biobased Polymers
ISBN 978-3-030-33896-1   ISBN 978-3-030-33897-8   (eBook)
https://doi.org/10.1007/978-3-030-33897-8

This Springer imprint is published by the registered company Springer Nature Switzerland AG
The registered company address is: Gewerbestrasse 11, 6330 Cham, Switzerland

Preface

The idea of writing a book on "Polyhydroxyalkanoates" struck me instantaneously after I completed my Ph.D. in biochemical engineering and biotechnology, and started working on polyhydroxyalkanoates from gram-negative bacteria during my postdoctoral work. After three years of research related to polyhydroxyalkanoates from crude glycerol, I decided to compile all the details as a small book on "Polyhydroxyalkanoates (PHAs)" and I found SpringerBrief series as the most appropriate way to publish our compendium. The prompt and positive response from the Springer editor team through their valuable suggestions and timely contributions is gratefully appreciated. This brief book consists of five chapters on different aspects; each one represents the progress on PHA types, synthesis, production strategies, and challenges in commercialization of PHA research and possible solutions. This is supposed to be the most up-to-date book on recent advances in polyhydroxyalkanoates. Attempt has been made not only to highlight the remarkable progress made by the scientific community in this field of research, but also to critically analyze the lacuna to expand the commercial prospects of these wonder biopolymers.

PHAs are exciting materials with their material properties quite similar to petrochemical-derived plastics such as polystyrene and polyethylene. The possibility of tailoring their composition and hence properties by regulation of appropriate growth and environmental conditions for PHA producers is truly fascinating. They have numerous potential applications, especially in medical and pharmaceutical industries such as tissue engineering, targeted drug delivery, and wound dressing. They are considered environmentally friendly since they are nontoxic and biodegradable. PHA biopolymers are biocompatible in nature which increases their potential for various biomedical applications. These features make PHAs advantageous in a wide variety of industrial applications. This book covers the current knowledge and the most recent advances in the field of microbial production of PHAs. This book includes the physicochemical properties of PHAs, different types of PHA, the strategies for their production using gram-negative and gram-positive bacteria, including the challenges in commercialization, and their possible biotechnological solutions.

First of all, I would like to express my profound gratitude to The Almighty God who has always been a source of my confidence, strength, and achievements. I thankfully acknowledge all my mentors for their valuable and inspiring contributions, throughout my research carrier, especially my postdoctoral mentors "Prof. Sanjeev K. Soni" and "Dr. Vijay Kumar Bharti" for giving me this opportunity. I do highly appreciate the help that I have constantly received from my Ph.D. mentor "Prof. Ashok Kumar Srivastava" at Indian Institute of Technology Delhi, New Delhi, India. I also thank my loving husband, Prince Sharma, and my mother, Smt. Nirmala Devi, for their constant support and love. I also thank my research fellows, particularly Nisha, Shubhangi, Armaan, Neha, Kalai, Preeti, Arup, and Jaspreet, for their support, understanding, and forbearance.

Chandigarh, India   Geeta Gahlawat

Contents

Abbreviations

3HB	3-hydroxybutyrate
3HDD	3-hydroxydodecanoate
3HHp	3-hydroxyheptanoate
3HHx	3-hydroxyhexanoate
3HO	3-hydroxyoctanoate
3HV	3-hydroxyvalerate
4HB	4-hydroxybutyrate
ADF	Aerobic dynamic feeding
ATPS	Aqueous two-phase system
CDW	Cell dry weight
CoA	Coenzyme A
COD	Chemical oxygen demand
CSTR	Continuous stirred-tank reactor
HA	Hydroxyalkanoate
HB	Hydroxybutyrate
ICI	Imperial Chemical Industries
LAS	Linear alkylbenzene sulfonic acid
LCL	Long-chain length
LPS	Lipo-polysaccharide
MCL	Medium-chain length
MIBK	Methyl isobutyl ketone
MMC	Mixed microbial culture
NPCM	Non-PHA cell material
PHA	Polyhydroxyalkanoate
PHB	Poly (3-hydroxybutyric acid)
Poly(3HB-co-3HHx)	Poly (3-hydroxybutyrate-co-3-hydroxyhexanoate)
Poly(3HB-co-3HV)	Poly (3-hydroxybutyrate-co-3-hydroxyvalerate)
Poly(3HB-co-3HV-co-3HHx)	Poly (3-hydroxybutyrate-co-3-hydroxyvalerate-co-3-hydroxyhexanoate)
Poly(3HB-co-4HB)	Poly (3-hydroxybutyrate-co-4-hydroxybutyrate)

Poly(3HHx-co-3HO)	Poly (3-hydroxyhexanoate-co-3-hydroxyoctanoate)
RBS	Ribosomal binding site
SBR	Sequencing batch reactor
SCL	Short-chain length
SDS	Sodium dodecyl sulfate
STR	Stirred-tank reactor
TPPB	Two-phase partitioning bioreactor
UHMW	Ultra-high molecular weight
v/v	Volume by volume
VA	Valeric acid
VFA	Volatile fatty acid
WWTP	Wastewater treatment plant

Symbols

a	Exponent indicating the type of relationship between S_2 (nitrogen) and μ
K_{S1}	Saturation constant for sucrose consumption (g/L)
K_{S2}	Saturation constant for nitrogen consumption (g/L)
$Y_{(X+P)S1}$	Yield of biomass and product on sucrose consumption (g/g)
$Y_{X/S2}$	Yield of biomass on nitrogen consumption (g/g)
m_{S1}	Maintenance coefficient for sucrose (g/g.h)
m_{S2}	Maintenance coefficient for nitrogen (g/g.h)
q_{S1}	Specific sucrose consumption rate (h^{-1})
q_{S2}	Specific nitrogen consumption rate (h^{-1})
q_p	Specific product formation rate (h^{-1})
D	Dilution rate (h^{-1})
F	Total flow rate (L/h)
F_1	Flow rate for sucrose (L/h)
F_2	Flow rate for nitrogen (L/h)
K_1	Growth-associated product formation constant (g/g)
K_I	Substrate inhibition constant for sucrose consumption (g/L)
n_1	Constant in equation
P	PHB concentration (g/L)
Qp	Volumetric productivity of PHB (g/L.h)
Qv	Volumetric productivity of P(3HB-co-3-HV) (g/L.h)
S_{01}	Inlet concentration of sucrose in the feed bottle (g/L)
S_{02}	Inlet concentration of nitrogen in the feed bottle (g/L)
S_1	Sucrose concentration (g/L)
S_2	Nitrogen concentration (g/L)
Sj	Variance of the error of a residual
S_m	Nitrogen concentration at which complete inhibition occurs (g/L)
V	Working volume of the bioreactor (L)

W_j	Weight of each variable
X	Biomass concentration (g/L)
dV/dt	Rate of change of volume (L/h)

Greek Symbols

μ_{max}	Maximum specific growth rate (h^{-1})
μ	Specific growth rate (h^{-1})
Δj	Mean residual of each variable (j)
Δ_{ij}	Difference between the model and the experimental values for ith data point and jth process variable

List of Figures

List of Tables

Chapter 1
Introduction and Background

Abstract The usage of synthetic plastics such as polyethylene and polypropylene was initiated by mankind to enhance the quality and comfort of life without realizing their ubiquitous nature. Now they have become an essential part of contemporary life and are being used increasingly in different industrial applications due to their unique characteristics of strength, durability and resistance to chemicals. The high molecular weight appears to be the main reason for the resistance of these plastics to biodegradation and perseverance in soil for a longer period of time. This non-biodegradable nature of synthetic plastics and dependency on fossil fuels for their production have driven the search for alternative sustainable biotechnological solution with lower environmental impact. In this regard, Polyhydroxyalkanoates (PHAs) are considered as best alternatives as they are produced by fermentation of renewable feedstock and are completely biodegradable. However, despite the considerable research work on PHAs, only limited success has been achieved so far. The main bottleneck in successful utilization of PHAs is their high cost of production. This book chapter presents general introduction on PHAs and their types, and how they came into existence.

Keywords Synthetic plastics · Polyhydroxyalkanoates · Sustainability · Renewable substrates · Classification · Copolymers

The synthetic plastics have continuously been used in several industrial and commercial applications due to their low cost and excellent properties of durability, resistance and good processability. They have been used in different applications such as home appliances, bottles, containers, sheets, aerospace materials, food packaging, and biomedical instrumentation and devices (Reddy et al. 2016). However, the use of synthetic plastics has created a problem world-wide because of their toxic, non-biodegradable nature and predicted end of global petroleum reserves in the near future (Loo and Sudesh 2007). The annual worldwide production of synthetic plastics has increased considerably over the last 60 years and they keep on accumulating in the environment at the rate of 25 million ton per year (Divya et al. 2013). Plastic materials take thousands of years to decompose in the environment and create huge problems for solid waste disposal (Cavalheiro et al. 2009). These plastics have a deadly impact on marine life through entanglement or direct ingestion of plastic

© The Author(s), under exclusive license to Springer Nature Switzerland AG 2019
G. Gahlawat, *Polyhydroxyalkanoates Biopolymers*,
Biobased Polymers, https://doi.org/10.1007/978-3-030-33897-8_1

materials or intoxication (Johnston 1990). Discarded plastics have adverse effects on the human health as well (Thompson et al. 2009). A variety of chemicals additives such as phthalates and bisphenol A (BPA) that are used in the plastics manufacturing are known to be toxic and cause reproductive abnormalities (Talsness et al. 2009). Another major concern which needs attention is that the plastic materials are non renewable in nature and are produced by the fractionation of depleting fossil fuels such as petrochemicals with the consumption of huge amounts of energy (Loo and Sudesh 2007). A recent survey done on the earth's mineral reserves demonstrated depletion of these valuable natural assets at faster rates. This problem has created a renewed interest to search for various other sustainable alternatives for the production of polymers.

In such a scenario, biopolymers are considered as best substitute for conventional plastics as they are derived from renewable feedstock and are biodegradable in the environment depending on the composition of polymer. Various types of biopolymers are under development that mainly includes polylactic acid, polyglycolic acid, aliphatic polyesters, polyhydroxyalkanoates (PHAs) and polysaccharides (Divya et al. 2013). Among all these, PHAs are considered as completely biodegradable, biocompatible and the most potential renewable alternative because of their almost similar mechanical and physical properties to that of conventional plastics (Akaraonye et al. 2010; Cavalheiro et al. 2012).

PHAs are polyesters of hydroxyalkanoic acids which are produced by various microbes intracellularly as energy and carbon storage materials in response to physiological stress conditions. Poly (3-hydroxybutyrate) or PHB is the most important member of PHAs family which has been studied widely. The importance of PHB is primarily due to its biocompatibility, easy degradation under both aerobic and/or anaerobic conditions and release of non-toxic products upon degradation. It has several promising applications in packaging, food, medical and pharmaceutical industries (Philip et al. 2007). However, PHB is still not able to substitute the synthetic polymers on the industrial scale because of its high production cost and low productivities (Wang et al. 2014). Moreover, it is highly brittle, stiff and thermally unstable during processing which further restricts its commercialization and market penetration (Koller et al. 2017). Introduction of other types of hydroxyl monomeric units i.e. 3-hydroxyvalerate (3HV) into 3-hydroxybutyrate (3HB) polymer produces a set of copolymers with improved mechanical properties such as tensile strength, elasticity and relatively quick degradation in vivo (García et al. 2013). It also decreases the brittleness and crystallinity of the polymer (Alsafadi and Al-Mashaqbeh 2017).

PHA copolymers are thermoplastic in nature and can be processed using conventional processing techniques which invariably extend their versatility in medical applications as well such as tissue engineering (Van-Thuoc et al. 2015). Poly (3-hydroxybutyrate-co-3-hydroxyvalerate) or P (3HB-co-3HV) copolymer is a well studied PHA which is produced by the addition of different precursors (propionate or valerate) to the medium along with the main carbon source. P (3HB-co-3HV) has much wider applications than PHB because of its low crystallinity and easier thermoprocessability (Wang et al. 2013). Hence, copolymer production can solve problems of stiffness and thermal unstability associated with PHB.

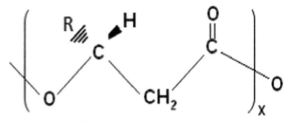

X varies from 100 to 30,000

R = hydrogen Poly (3-hydroxypropionate);   R = methyl   Poly (3-hydroxybutyrate)

R = ethyl   Poly (3-hydroxyvalerate);   R = propyl   Poly (3-hydroxyhexanoate)

R = pentyl   Poly (3-hydroxyoctanoate)

Fig. 1.1 General structure of PHAs (Verlinden et al. 2007; Khanna and Srivastava 2005)

Polyhydroxyalkanoates (PHAs)—An Alternative to Synthetic Plastics

Polyhydroxyalkanoates or PHAs are bacterial polyesters which are composed of 3-hydroxyalkanoic acids (HA acids) (structure shown in Fig. 1.1). They are synthesized and accumulated by various bacteria as an intracellular energy storage material under the nutrients limitation such as nitrogen, phosphorous, or sulfur with excess carbon source availability. PHAs are excellent reserve materials because their intracellular nature does not affect the osmotic pressure of the cell (Loo and Sudesh 2007). PHAs have received a great deal of attention as promising alternative for synthetic plastics because of their similar material properties such as tensile strength and melting point to petroleum-derived plastics. Moreover, these polymers can be produced from renewable resources such as vegetable oils, sugars, molasses and syngas, and therefore can help in reducing the dependency of society on rapidly depleting non-renewable fossil-fuel resources for synthetic polymer production (Loo and Sudesh 2007).

At present, more than 150 different monomers (HA units) can be combined within PHAs family to yield materials (homo-polymers or copolymers) with extremely different physical properties (Steinbüchel and Eversloh 2003). The side chain length and functional groups greatly affects the physicochemical properties of the PHAs polymers such as glass transition temperature, crystallinity and melting point which in turn decide their final usage. Depending upon the type of carbon sources (substrates) fed to the microorganisms, cultivation conditions, bacterial strains and metabolic pathways active inside the cells, homo- or copolyesters with different HAs units could be generated or accumulated in the cells (Sudesh 2000).

1.1   Emergence of PHAs and Their Characteristics

The homopolymer PHB is the first member of the PHAs group to be discovered by Maurice Lemoigne in 1926 (Lemoigne 1926). By the end of 1950s, reports on the *Bacillus* demonstrated that PHB functions as an intracellular energy and carbon storage material in these bacteria (Macrae and Wilkinson 1958; Williamson and Wilkinson 1958). Later on, it was found that the synthesis of this intracellular storage polymer is a popular phenomenon in gram-negative bacteria (Forsyth et al. 1958). In 1970s, Wallen and Rohwedder (1974) first time reported the existence of 3HA units other than the 3-hydroxybutyrate (3HB) in PHAs. Among the 3HA units, 3HV and 3-hydroxyhexanoate (3HHx) were the major and minor units, respectively found in the synthesized polymer from activated sewage sludge. Synthesis of PHAs, in the form of granules, inside the cytoplasm was also reported by many gram-positive bacteria (Findlay and White 1983). In this report it was demonstrated that *Bacillus megaterium* could accumulate PHAs containing 95% of 3HB, 3% of 3-hydroxyheptanoate (3HHp), 2% of an 8-carbon HA and traces of other HA units. De Smet and co-workers (1983) have grown *Pseudomonas oleovorans* on *n*-octane and the composition analysis of polymer demonstrated the presence of chiefly 3-hydroxyoctanoate (3HO) and 3-hydroxyhexanoate (3HHx) units. These findings of HA units other than the 3HB units were the major landmarks in the beginning of another area of research on PHAs other than the PHB.

   PHAs are linear polyesters of β-hydroxyalkanoates wherein monomeric units exist in [R]-configuration (Fig. 1.1). The main reason for this is the stereo-specificity of the polymerizing enzyme, PHA synthase. PHAs are partially crystalline polymers with crystallinity in the range of 60–80%. The molecular mass of these polymers ranges from 200 to 3000 KDa, depending upon the substrate used, microorganism and cultivation conditions (Divya et al. 2013). They exist as amorphous granules inside the cells and appear as highly refractive granules under electron microscopic observations (Sudesh et al. 2000). Rapid crystallization of PHAs was observed as polymer granules were extracted from the cells. This can be described by a kinetic nucleation phenomenon. The polymer granules within the cells exist in the form of small granules and probability of a nucleation event triggering crystallization is very low (Khanna and Srivastava 2005). But as the cells are disrupted, granules starts coalescing and quick heterogeneous nucleation occurs. Microorganisms accumulating PHAs can be easily identified by staining with Sudan Black B or Nile Blue A dye (Burdon et al. 1942; Williamson and Wilkinson 1958). But, PHAs are more specifically stained by the fluorescent dye Nile Blue A and visualized by epifluorescence microscope under oil immersion (Ostle and Holt 1982).

   The physico-mechanical properties of PHA are very much dependent upon its monomeric unit composition, therefore biopolymers having tailor-made properties can be designed by introducing different monomeric units (Sudesh and Doi 2005). PHAs have various other remarkable properties such as piezoelectric activity, nonlinear optical activity and biocompatibility by virtue of which they have found place in other medical areas such as drug delivery, bone repair and cartilage tissue repair

(William and Martin 2002). Hydrolytic degradation of PHAs takes place by surface erosion, making it an ideal material for controlled drug release application.

1.2 Classification of PHAs

PHAs can be divided into two groups on the basis of number of carbon atoms present in the monomer units (Sudesh et al. 2000). The first group comprises of short chain length (SCL) polymers containing 3–5 carbon atoms such as PHB and poly (3-hydroxybutyrate-co-3-hydroxyvalerate) copolymer, synthesized by various bacteria such as *Cupriavidus necator* (earlier known as *Ralstonia eutropha*) and *Alcaligenes latus* (Akaraonye et al. 2010). The second group includes medium chain length (MCL) polymers containing 6–14 carbon atoms such as poly (3-hydroxyhexanoate-co-3-hydroxyoctanoate) or poly (3HHx-co-3HO), synthesized by *Pseudomonas putida*, *Pseudomonas aeruginosa* and *Pseudomonas oleovorans*. PHAs having more than 14 carbon atoms are considered as long chain length PHAs (LCL-PHAs) and, they are very rare and least studied (Singh and Mallick 2008). Specificity of PHA synthases towards different substrates help in determining the carbon chain length of 3HAs monomers.

1.2.1 Polyhydroxybutyrate (PHB)

PHB is the most well known member of PHAs family of polyester biopolymers. This polymer has gained much biotechnological interest because it offers the possibility for renewable resource-based biodegradable plastic materials (Zinn et al. 2001; Panda et al. 2006). It is an isotactic polymer which is rather stiff and highly crystalline in nature. It is accumulated as intracellular inclusions inside the cell up to 90% of the cell dry weight. Mechanical properties of PHB are similar to that of polypropylene and polystyrene, however elongation to break is significantly lower than their counterparts. PHB is a highly stiff material which becomes brittle upon storage at room temperature conditions. The melting temperature (175 °C) of PHB is quite closer to its degradation temperature (220 °C) which makes thermal processing a difficult task (Tokiwa and Ugwu 2007). PHB has various interesting properties such as good resistance to moisture, UV stability, impermeability to oxygen and optical purity which makes it suitable for packaging applications (Ebnesajjad 2012). On the contrary, other biodegradable polymers are either soluble in water or sensitive to moisture which limits their application in packaging. The accumulation of PHB inside the cells of *Alcaligenes eutrophus* under nitrogen limitation leads to an increase in buoyant density of the cells (Pedrós Alió et al. 1985). As the PHB content increases from 0 to 1.699 pg cell^{-1}, a three times increase in volume of the cells of *A. eutrophus* strain N9A was observed.

1.2.2   Poly (3-hydroxybutyrate-co-3-hydroxyvalerate) [Poly (3HB-co-3HV)]

Poly(3HB-co-3HV) is a copolymer of 3HB and 3HV which have been studied widely because of its toughness and excellent processability. Increased elongation to break and low melting temperature range of 100–170 °C (depending upon the 3HV fraction) does not degrade it during thermal processing at higher temperature, and further enhances its application potential in diverse areas (Tokiwa and Ugwu 2007). Incorporation of 3HV monomeric units into a polymer consisting mainly of 3HB leads to a decrease in crystallinity and melting temperature compared to PHB homopolymer. This translates to decrease in rigidity and an increase in the tensile strength of polymer particularly in terms of its mechanical strength (Loo and Sudesh 2007).

The mechanical properties of poly (3HB-co-3HV) are reported to be mainly dependent on the molar ratio of 3HV. Poly (3HB-co-3HV) became more flexible with an increase of 3HV content from 0 to 25 mol%, which also prevents its degradation during thermal processing (Holmes 1985; Doi 1990). Poly (3HB-co-3HV) copolymer was successfully synthesized and marketed under the trade name of BIOPOL by Imperial Chemical Industries (ICI) in 1981 by employing bacteria, *R. eutropha*. During fermentation, propionate and glucose were used as 3HV and 3HB-precursors, respectively (Verlinden et al. 2007). Although propionic acid was used as 3HV-precursor but it had two shortcomings: firstly, it was very expensive and did not exclusively generate 3HV units. Secondly, high concentration of propionic acid was highly toxic and caused cell death. Therefore it should be added at low concentrations to avoid the cell damage (Doi 1990; Yu et al. 1999).

1.2.3   Poly (3-hydroxybutyrate-co-3-hydroxyhexanoate) [Poly (3HB-co-3HHx)]

Poly (3HB-co-3HHx) is an exciting copolymer because this polymer is a combination of SCL monomer units, 3HB (highly crystalline) and MCL monomer units, 3HHx (elastomeric) which makes it suitable for tissue engineering (Loo and Sudesh 2007). The mechanical and processing properties of poly (3HB-co-3HHx) have been observed to be better than PHB and poly (3HB-co-3HV) that are commercially available (Doi et al. 1995). PHB is very fragile which restricts its usage in cartilage engineering whereas blending of PHB with 3HHx improves its mechanical properties and gives it better flexibility, thereby making it a very useful material in cartilage tissue engineering (Wang et al. 2008). Poly (3HB-co-3HHx) has proved to be a promising medical implant biomaterial due to its biocompatibility, resorbability, non-toxicity and better elastomeric properties (Zhao et al. 2003). Introduction of 10 mol% of 3HHx monomer units into the 3HB unit enhances the elongation to break from 5 to 400%, thereby resulting in more flexible polymer (Sudesh et al. 2000). Moreover, introduction of even small quantities of 3HHx units into the PHB polymer decreases the melting temperature from 180 to 155 °C, hence significantly improves the thermal processability and physico-mechanical properties of the polymer (Loo et al. 2005; Doi et al. 1995).

1.3 Biochemical Synthesis of PHAs

Since 1987, the extensive literature survey has been done on PHAs biosynthesis metabolic pathways and biochemistry of their enzymes (Madison and Huisman 1999; Reinecke and Steinbuechel 2008). Various genes coding for PHA synthesis and degradation enzymes have been cloned and studied in detail in literature. These reports have indicated that the microorganisms have evolved various different metabolic pathways for PHA synthesis. Moreover, the number of metabolic pathways to PHAs synthesis is increasing due to genetic engineering approaches and expression of PHA synthesis enzymes in foreign hosts (Zinn et al. 2001). The different pathways of PHAs biosynthesis found in various bacteria are summarized in the following sections:

1.3.1 SCL-PHAs Synthesis [PHB and Poly (3HB-co-3HV)]

PHB synthesis is a widely distributed phenomenon in bacteria but the biochemical investigations on the metabolic pathway of PHB synthesis has extensively been focused on mainly *C. necator* (Reinecke and Steinbuechel 2008). When a carbon source is supplied such as glucose and fructose sugars, biosynthetic pathway leads to synthesis of PHB via pathways I (as shown in Fig. 1.2a). On the other hand PHAs copolymers are produced when sugars and fatty acids are metabolized by the pathway II and III (Verlinden 2007; Tsuge 2002). Pathway II follows the fatty acid degradation route whereas Pathway III leads to fatty acid synthesis and shown in Figs. 1.2b and 1.3.

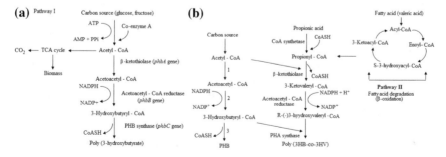

Fig. 1.2 a Metabolic pathway of SCL-PHA synthesis in *C. necator* [adapted and modified from Lee et al. (1999)]; **b** Metabolic pathway of poly (3HB-co-3HV) synthesis [adapted from Loo and Sudesh (2007)]. Enzymes: 1-β-ketothiolase; 2-acetoacetyl-CoA reductase; 3-PHB synthase

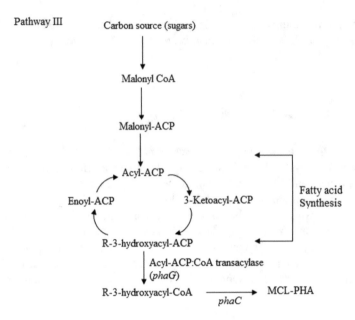

Fig. 1.3 Metabolic pathway of MCL-PHA synthesis in *Pseudomonas* from carbohydrates (adapted from Madison and Huisman 1999)

The synthesis of PHB involves three different enzymes mainly β-ketothiolase, NAD(P)H-dependent acetoacetyl-CoA reductase and PHB synthase which are coded by *phbA*, *phbB*, *phbC* genes respectively (Doi 1990; Anderson and Dawes 1990). In first step, β-ketothiolase condenses the two molecules of acetyl-CoA into acetoacetyl-CoA which is then converted into 3-hydroxybutyryl-CoA using NAD(P)H-dependent acetoacetyl-CoA reductase enzyme. Finally, 3-hydroxybutyryl-CoA units are polymerized into poly(3-hydroxybutyrate) using PHB synthase as shown in Fig. 1.2a (Tsuge 2002). *C. necator* expresses or synthesizes two types of β-ketothiolases: enzyme A and B. β-ketothiolases A coded by *phbA* gene is involved in the synthesis of PHB whereas β-ketothiolases B in PHAs copolymer synthesis (Slater et al. 1998). The main difference between these two β-ketothiolases is their specificity for different substrates. It has been reported that enzyme β-ketothiolases B efficiently utilize 3-ketovaleryl-CoA as 3HV monomer unit for poly (3HB-co-3HV) synthesis.

PHAs with different monomer composition could be synthesized by *C. necator* and, the type and composition of monomeric units depends upon the nature of carbon sources fed during the cultivation. Supplementation of propionate or valerate to the growth medium having glucose/fructose results in the production of poly(3HB-co-3HV) copolymer composed of 3HB and 3HV units (Fig. 1.2b). Besides 3-hydroxybutyryl-CoA, biosynthesis of poly(3HB-co-3HV) requires 3-hydroxyvaleryl-CoA (3HV-CoA) unit. The 3HV-CoA unit is synthesized by the condensation of propionyl-CoA and acetyl-CoA into 3-ketovaleryl-CoA unit which is subsequently converted to 3HV-CoA unit by the β-ketothiolases B (Steinbüchel and Eversloh 2003). Various other propionigenic fatty acid substrates having odd number of carbon such as valeric acid and heptanoic acid etc. could also serve as precursors for poly (3HB-co-3HV) synthesis. The β-oxidation of these fatty acids via pathway II results in propionyl-CoA moieties (as shown in Fig. 1.2b) which can directly combine with acetyl CoA and synthesize 3-hydroxyvaleryl-CoA monomer units for copolymer synthesis (Steinbüchel and Eversloh 2003).

1.3.2 MCL-PHAs Synthesis in Pseudomonas *sp.*

Pseudomonas species falling under the category of rRNA homology group I (except *Pseudomonas oleovorans*) synthesize medium chain length PHAs containing C_8 to C_{10} monomers when they are grown on carbohydrates (Haywood et al. 1990). It has been reported that these monomeric units are obtained from the intermediates of fatty acid de novo metabolic pathway (as shown in Fig. 1.3, Pathway III). This is the main pathway involved in the synthesis of 3-hydroxyacyl moieties which are the major constituents (approximately 90%) of the accumulated MCL-PHA in *P. putida*, *P. aeruginosa* and other pseudomonads (Huijberts et al. 1992, 1994; Hoffmann et al. 2000a). The conversion of (R)-3-hydroxyacyl-ACP intermediate to (R)-3-hydroxyacyl-CoA is catalyzed by the enzyme, 3-hydroxyacyl-CoA-ACP transacylase encode by *phaG* gene (Fig. 1.3; Rehm et al. 1998). The key enzyme, 3-hydroxyacyl-CoA-ACP transacylase links the both fatty acid synthesis pathway and MCL-PHA biosynthesis pathway in pseudomonads. In *P. oleovorans* the *phaG* gene coding for the 3-hydroxyacyl-CoA-ACP transacylase is silent describing why this species is incapable of synthesizing MCL-PHAs from carbohydrates (Hoffmann et al. 2000b; Rehm et al. 2001).

Most *Pseudomonas* sp. including *P. oleovorans* follows a different metabolic pathway for MCL-PHA synthesis when grown on alkanes, alkanols and alkanoates which is popularly known as MCL-PHA biosynthetic pathway. In this metabolic pathway, 3-hydroxyacyl-CoA intermediates obtained from fatty acid β-oxidation are channeled towards PHA synthesis (Huisman et al. 1989). The fatty acids are generally metabolized by the removal of acetyl-CoA units and, acyl-CoA thus formed gets oxidized to 3-ketoacyl-CoA via (S)-3-hydroxyacyl-CoA intermediate (Fig. 1.2b). Intermediates of β-oxidation pathway are then converted to (R)-3-hydroxyacyl-CoA by three enzymes namely hydratase, epimerase and 3-ketoacyl-CoA reductase (Lageveen et al. 1988; Timm and Steinbüchel 1990).

1.3.3 Other Pathways Involved in PHAs Synthesis

Other bacteria such as *Aeromonas hydrophila* and *Aeromonas caviae* are capable of synthesizing MCL-PHA, a copolymer of 3HB and 3HHx when grown on even-number fatty acids such as olive oil. On the other hand, growth on odd-numbered

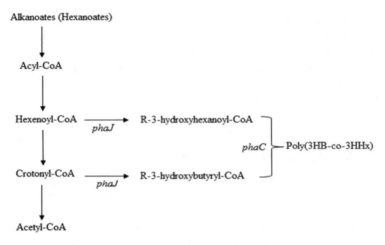

Fig. 1.4 Metabolic pathway of poly (3HB-co-3HHx) synthesis in *Aeromonas caviae*. Adapted and modified from Madison and Huisman (1999)

fatty acids results in poly (3HB-co-3HV) copolymers that consist of primarily 3HV and small amounts of 3HB units (Doi et al. 1995). The PHA gene locus in *A. caviae* encodes three different enzymes namely PHA polymerase (*phaC*), enoyl-CoA hydratase (*phaJ*) and a phasin (*phaP*) (Fukui and Doi 1997; Fukui et al. 1998). A (R)-specific enoyl-CoA hydratase encoded by *phaJ* has indicated that PHA biosynthetic pathway in *A. caviae* starts from enoyl-CoA derivative and this enzyme help in the bioconversion of trans-2-enoyl-CoA to 3-hydroxyacyl-CoA when grown on hexanoate. Figure 1.4 illustrates the complete metabolic pathway of poly (3HB-co-3HHx) synthesis in *A. caviae*. Apart from conversion of crotonyl-CoA to 3-hydroxybutyryl-CoA, *PhaJ* also help in conversion of pentenoyl-CoA and hexenoyl-CoA into PHA precursors (Madison and Huisman 1999; Fukui et al. 1999).

Some bacteria such as *Rhodococcus ruber* and *Nocardia corallina* are able to synthesize PHAs copolymers containing 3HV units without the use of typical 3HV precursors such as valerate and propionate in the growth medium (Valentin and Dennis 1996). In this pathway, 3HV unit is obtained from acetyl-CoA and propionyl-CoA, where the latter is synthesized by methylmalonyl-CoA pathway not by common route of poly (3HB-co-3HV) synthesis as mentioned in Sect. 1.3.1. It has been reported that in this pathways, succinyl-CoA is reduced to methylmalonyl-CoA which is then decarboxylated to propionyl-CoA (Williams et al. 1994).

1.4 Regulation of PHAs Synthesis

The accumulation of PHA inside the cell is directly dependent on the intracellular concentrations of NADH and acetyl-CoA molecules. The key step in the regulation of PHB synthesis is the fate of acetyl-CoA which either can enter into tricarboxylic acid cycle or TCA cycle or can serve as the precursor for PHB biosynthesis (Byrom 1987; Sudesh et al. 2000). The fate of acetyl-CoA moiety depends upon the environmental conditions such as nutrient limitation or balanced growth conditions. At the time of nutrient limitation, intracellular concentrations of NADH increases which further inhibit the main regulatory enzymes of TCA cycle i.e. citrate synthase and isocitrate dehydrogenase resulting in down regulation of TCA cycle. As a result, acetyl-CoA does not take part in TCA cycle and gets accumulated inside the cells. This continuous accumulation of acetyl CoA is concomitant with the decrease in free Co-A level which activates the β-ketothiolase enzyme of the PHB pathway and starts PHB synthesis (Page and Knosp 1989). Under balanced growth conditions in the absence of nutrient limitation, acetyl-CoA units enter into TCA cycle and resultant free CoA level starts increasing. High concentrations of free CoA units then inhibit the β-ketothiolase enzyme and thereby prevent the synthesis of PHB. It has been reported that PHA synthesizing bacteria use PHA not only as energy reserve material but also as a sink for reduction equivalents, and hence play the role of redox regulator (Senior and Dawes 1971; Genser et al. 1998). Some bacteria such as *A. latus* (Hrabak 1992) and a mutant strain of *Azotobacter vinelandii* (Page and Knosp 1989) are capable of accumulating PHAs during growth phase of cultivation without any nutrient limitation. However, the use of nutrient limitation in *A. latus* culture was demonstrated to further enhance the accumulation of PHAs (Lee et al. 1999)

Attempt has been made in present book to summarize the current advances in the biotechnological production of PHAs, and the challenges in their mass scale production and commercialization. The production protocols of PHAs by different bacteria using renewable carbon substrates and process optimization including high cell density cultivation strategies have been summarized in this book. Finally efficient product recovery protocols and future perspectives of PHAs have been discussed in detail in order to outline the progress made so far in the area of economic biopolymer production.

References

Akaraonye E, Keshavarz T, Roy I (2010) Production of polyhydroxyalkanoates: the future green materials of choice. J Chem Technol Biotechnol 85(6):732–743

Alsafadi D, Al-Mashaqbeh O (2017) A one-stage cultivation process for the production of poly-3-(hydroxybutyrate-co-hydroxyvalerate) from olive mill wastewater by *Haloferax mediterranei*. New Biotechnol 34:47–53

Anderson AJ, Dawes EA (1990) Occurrence, metabolism, metabolic role, and industrial uses of bacterial polyhydroxyalkanoates. Microbiol Rev 54(4):450–472

Burdon KL, Stokes JC, Kimbrough CE (1942) Studies of the common aerobic spore-forming bacilli: I. Staining for fat with Sudan black B-safranin. J Bacteriol 43(6):717

Byrom D (1987) Polymer synthesis by microorganisms: technology and economics. Trends Biotechnol 5(9):246–250

Cavalheiro JMBT, de Almeida MCMD, Grandfils C, da Fonseca MMR (2009) Poly(3-hydroxybutyrate) production by *Cupriavidus necator* using waste glycerol. Process Biochem 44(5):509–515

Cavalheiro JMBT et al (2012) Effect of cultivation parameters on the production of poly(3-hydroxybutyrate-co-4-hydroxybutyrate) and poly(3-hydroxybutyrate-4-hydroxybutyrate-3-hydroxyvalerate) by *Cupriavidus necator* using waste glycerol. Bioresour Technol 111:391–397

De Smet M et al (1983) Characterization of intracellular inclusions formed by *Pseudomonas oleovorans* during growth on octane. J Bacteriol 154(2):870–878

Divya G, Archana T, Manzano RA (2013) Polyhydroxy alkonates—a sustainable alternative to petro-based plastics. J Petrol Environ Biotechnol 4:1–8

Doi Y (1990) Microbial polyesters. VCH Publishers, New York

Doi Y, Kitamura S, Abe H (1995) Microbial synthesis and characterization of poly (3-hydroxybutyrate-co-3-hydroxyhexanoate). Macromolecules 28(14):4822–4828

Ebnesajjad S (2012) Plastic films in food packaging: materials, technology and applications. Elsevier William Andrew Publishers, Oxford

Findlay RH, White DC (1983) Polymeric beta-hydroxyalkanoates from environmental samples and *Bacillus megaterium*. Appl Environ Microbiol 45(1):71–78

Forsyth W, Hayward A, Roberts J (1958) Occurrence of poly-β-hydroxybutyric acid in aerobic gram-negative bacteria. Nature 182(4638):800–801

Fukui T, Doi Y (1997) Cloning and analysis of the poly (3-hydroxybutyrate-co-3-hydroxyhexanoate) biosynthesis genes of *Aeromonas caviae*. J Bacteriol 179(15):4821–4830

Fukui T, Shiomi N, Doi Y (1998) Expression and characterization of (R)-specific enoyl coenzyme a hydratase involved in polyhydroxyalkanoate biosynthesis by *Aeromonas caviae*. J Bacteriol 180(3):667–673

Fukui T, Yokomizo S, Kobayashi G (1999) Co-expression of polyhydroxyalkanoate synthase and (R)-enoyl-CoA hydratase genes of *Aeromonas caviae* establishes copolyester biosynthesis pathway in escherichia coli. FEMS Microbiol Lett 170(1):69–75

García IL et al (2013) Evaluation of by-products from the biodiesel industry as fermentation feedstock for poly(3-hydroxybutyrate-co-3-hydroxyvalerate) production by *Cupriavidus necator*. Bioresour Technol 130:16–22

Genser KF, Renner G, Schwab H (1998) Molecular cloning, sequencing and expression in *Escherichia coli* of the poly(3-hydroxyalkanoate) synthesis genes from *Alcaligenes latus* DSM1124. J Biotechnol 64(2–3):123–135

Haywood GW, Anderson AJ, Ewing DF, Dawes EA (1990) Accumulation of a polyhydroxyalkanoate containing primarily 3-hydroxydecanoate from simple carbohydrate substrates by Pseudomonas sp. strain NCIMB 40135. Appl Environ Microbiol 56(11):3354–3359

Hoffmann N, Steinbüchel A, Rehm BH (2000a) Homologous functional expression of cryptic phaG from *Pseudomonas oleovorans* establishes the transacylase-mediated polyhydroxyalkanoate biosynthetic pathway. Appl Microbiol Biotechnol 54(5):665–670

Hoffmann N, Steinbüchel A, Rehm BH (2000b) The *Pseudomonas aeruginosa* phaG gene product is involved in the synthesis of polyhydroxyalkanoic acid consisting of medium-chain-length constituents from non-related carbon sources. FEMS Microbiol Lett 184(2):253–259

Holmes P (1985) Applications of PHB-a microbially produced biodegradable thermoplastic. Phys Technol 16(1):32

Hrabak O (1992) Industrial production of poly-β-hydroxybutyrate. FEMS Microbiol Lett 103(2–4):251–255

Huijberts G, de Rijk TC, de Waard P, Eggink G (1994) 13C nuclear magnetic resonance studies of *Pseudomonas putida* fatty acid metabolic routes involved in poly (3-hydroxyalkanoate) synthesis. J Bacteriol 176(6):1661–1666

Huijberts G, Eggink G, De Waard P, Huisman GW, Witholt B (1992) *Pseudomonas putida* KT2442 cultivated on glucose accumulates poly (3-hydroxyalkanoates) consisting of saturated and unsaturated monomers. Appl Environ Microbiol 58(2):536–544

Huisman GW, de Leeuw O, Eggink G, Witholt B (1989) Synthesis of poly-3-hydroxyalkanoates is a common feature of fluorescent pseudomonads. Appl Environ Microbiol 55(8):1949–1954

Johnstone B (1990) A throw away answer. Far East Econ Rev 147:62–63

Khanna S, Srivastava AK (2005) Recent advances in microbial polyhydroxyalkanoates. Process Biochem 40(2):607–619

Koller M, Maršálek L, de Sousa Dias MM, Braunegg G (2017) Producing microbial polyhydroxyalkanoate (PHA) biopolyesters in a sustainable manner. New Biotechnol 37:24–38

Lageveen RG, Huisman GW, Preusting H, Ketelaar P, Eggink G and Witholt B (1988) Formation of Polyesters by *Pseudomonas oleovorans*: Effect of Substrates on Formation and Composition of Poly-(R)-3-Hydroxyalkanoates and Poly-(R)-3-Hydroxyalkenoates. Appl Environ Microbiol 54(12):2924–2932

Lemoigne M (1926) Products of dehydration and of polymerization of β-hydroxybutyric acid. Bull Soc Chem Biol 8:770–782

Lee SY, Lee Y, Wang F (1999) Chiral compounds from bacterial polyesters: sugars to plastics to fine chemicals. Biotechnol Bioeng 65(3):363–368

Loo CY, Sudesh K (2007) Polyhydroxyalkanoates: bio-based microbial plastics and their properties. Malays Polym J 2(2):31–57

Loo CY et al (2005) Biosynthesis and characterization of poly (3-hydroxybutyrate-co-3-hydroxyhexanoate) from palm oil products in a *Wautersia eutropha* mutant. Biotechnol Lett 27(18):1405–1410

Macrae R, Wilkinson J (1958) Poly-β-hyroxybutyrate metabolism in washed suspensions of *Bacillus cereus* and *Bacillus megaterium*. J Gen Appl Microbiol 19(1):210–222

Madison LL, Huisman GW (1999) Metabolic Engineering of Poly(3-Hydroxyalkanoates): From DNA to Plastic. Microbiol Mol Biol Rev 63(1):21–53

Ostle AG, Holt J (1982) Nile blue A as a fluorescent stain for poly-beta-hydroxybutyrate. Appl Environ Microbiol 44(1):238–241

Page WJ, Knosp O (1989) Hyperproduction of poly-β-hydroxybutyrate during exponential growth of *Azotobacter vinelandii* UWD. Appl Environ Microbiol 55(6):1334–1339

Panda B, Jain P, Sharma L, Mallick N (2006) Optimization of cultural and nutritional conditions for accumulation of poly-β-hydroxybutyrate in *Synechocystis* sp. PCC 6803. Bioresour Technol 97(11):1296–1301

Pedrós-Alió C, Mas J, Guerrero R (1985) The influence of poly-β-hydroxybutyrate accumulation on cell volume and buoyant density in *Alcaligenes eutrophus*. Arch Microbiol 143(2):178–184

Philip S, Keshavarz T, Roy I (2007) Polyhydroxyalkanoates: biodegradable polymers with a range of applications. J Chem Technol Biotechnol 82(3):233–247

Reddy MV et al (2016) Production of poly-3-hydroxybutyrate (P3HB) and poly (3-hydroxybutyrate-co-3-hydroxyvalerate) P (3HB-co-3HV) from synthetic wastewater using *Hydrogenophaga palleronii*. Bioresour Technol 215:155–162

Rehm BH, Krüger N, Steinbüchel A (1998) A New Metabolic Link between Fatty Acid de NovoSynthesis and Polyhydroxyalkanoic Acid Synthesis The PHAG Gene from *Pseudomonas Putida* kt2440 Encodes A 3-Hydroxyacyl-Acyl Carrier Protein-Coenzyme A Transferase. J Biol Chem 273(37):24044–24051

Rehm BH, Mitsky TA and Steinbüchel A (2001) Role of Fatty Acid De Novo Biosynthesis in Polyhydroxyalkanoic Acid (PHA) and Rhamnolipid Synthesis by Pseudomonads: Establishment of the Transacylase (PhaG)-Mediated Pathway for PHA Biosynthesis in Escherichia coli. Appl Environ Microbiol 67(7):3102–3109

Reinecke F, Steinbuechel A (2008) *Ralstonia eutropha* strain H16 as model organism for PHA metabolism and for biotecAdd hnological production of technically interesting biopolymers. J Mol Microbiol Biotechnol 16(1–2):91–108

Senior P, Dawes E (1971) Poly-beta-hydroxybutyrate biosynthesis and the regulation of glucose metabolism in *Azotobacter beijerinckii*. Biochem J 125:55–66

Singh AK, Mallick N (2008) Enhanced production of SCL-LCL-PHA co-polymer by sludge-isolated *Pseudomonas aeruginosa* MTCC 7925. Lett Appl Microbiol 46(3):350–357

Slater S, Houmiel KL, Tran M, Mitsky TA, Taylor NB, Padgette SR, Gruys KJ (1998) Multiple β-ketothiolases mediate poly (β-hydroxyalkanoate) copolymer synthesis in Ralstonia eutropha. J Bacteriol 180(8):1979–1987

Steinbüchel A, Lütke-Eversloh T (2003) Metabolic engineering and pathway construction for biotechnological production of relevant polyhydroxyalkanoates in microorganisms. Biochemical Eng J 16(2):81–96

Sudesh K (2000) Molecular design and biosynthesis of biodegradable polyesters. Polym Adv Technol 11(8–12):865–872

Sudesh K, Abe H, Doi Y (2000) Synthesis, structure and properties of polyhydroxyalkanoates: biological polyesters. Progress Polym Sci 25(10):1503–1555

Sudesh K, Doi Y (2005) Polyhydroxyalkanoates. Handbook of biodegradable polymers, pp 219–256

Talsness CE, Andrade AJ, Kuriyama SN, Taylor JA, vom Saal FS (2009) Components of plastic: experimental studies in animals and relevance for human health. Philosophical Trans Royal Soc B: Biological Sci 364(1526):2079–2096

Thompson RC, Moore CJ, vom Saal FS, Swan SH (2009) Plastics, the environment and human health: current consensus and future trends. Philosophical Trans Royal Soc B: Biol Sci 364(1526):2153–2166

Timm A, Steinbüchel A (1990) Formation of polyesters consisting of medium-chain-length 3-hydroxyalkanoic acids from gluconate by Pseudomonas aeruginosa and other fluorescent pseudomonads. Appl Environ Microbiol 56(11):3360–3367

Tokiwa Y, Ugwu CU (2007) Biotechnological production of (R)-3-hydroxybutyric acid monomer. J Biotechnol 132(3):264–272

Tsuge T (2002) Metabolic improvements and use of inexpensive carbon sources in microbial production of polyhydroxyalkanoates. J Biosci Bioeng 94(6):579–584

Valentin H, Dennis D (1996) Metabolic pathway for poly (3-hydroxybutyrate-co-3-hydroxyvalerate) formation in Nocardia corallina: inactivation of mutB by chromosomal integration of a kanamycin resistance gene. Appl Environ Microbiol 62(2):372–379

Van-Thuoc D, Huu-Phong T, Minh-Khuong D, Hatti-Kaul R (2015) Poly (3-hydroxybutyrate-co-3-hydroxyvalerate) production by a moderate halophile Yangia sp. ND199 using glycerol as a carbon source. Appl Biochem Biotechnol 175(6):3120–3132

Verlinden RA et al (2007) Bacterial synthesis of biodegradable polyhydroxyalkanoates. J Appl Microbiol 102(6):1437–1449

Verlinden RA, Hill DJ, Kenward M, Williams CD, Radecka I (2007) Bacterial synthesis of biodegradable polyhydroxyalkanoates. J Appl Microbiol 102(6):1437–1449

Wallen LL, Rohwedder WK (1974) Poly-β-hydroxyalkanoate from activated sludge. Environ Sci Technol 8(6):576–579

Wang Y, Bian YZ, Wu Q, Chen GQ (2008) Evaluation of three-dimensional scaffolds prepared from poly(3-hydroxybutyrate-co-3-hydroxyhexanoate) for growth of allogeneic chondrocytes for cartilage repair in rabbits. Biomaterials 29(19):2858–2868

Wang Y et al (2013) Biosynthesis and thermal properties of PHBV produced from levulinic acid by Ralstonia eutropha. PLoS ONE 8(4):1–8

Wang Y, Yin J, Chen GQ (2014) Polyhydroxyalkanoates, challenges and opportunities. Curr Opin Biotechnol 30:59–65

Williams SF, Martin DP (2002) Applications of PHAs in medicine and pharmacy. Biopolymers 4:91–127

Williamson D, Wilkinson J (1958) The isolation and estimation of the poly-β-hydroxybutyrate inclusions of Bacillus Species. J Gen Appl Microbiol 19(1):198–209

Williams DR, Anderson AJ, Dawes EA, Ewing DF (1994) Production of a co-polyester of 3-hydroxybutyric acid and 3-hydroxyvaleric acid from succinic acid by Rhodococcus ruber: biosynthetic considerations. Appl Microbiol Biotechnol 40(5):717–723

Yu PH, Chua H, Huang AL, Ho KP (1999) Conversion of industrial food wastes by Alcaligenes latus into polyhydroxyalkanoates. Appl Biochem Biotechnol 78(1–3):445–454

Zhao K, Deng Y, Chun CJ, Chen GQ (2003) Polyhydroxyalkanoate (PHA) scaffolds with good mechanical properties and biocompatibility. Biomaterials 24(6):1041–1045

Zinn M, Witholt B, Egli T (2001) Occurrence, synthesis and medical application of bacterial polyhydroxyalkanoate. Adv Drug Deliv Rev 53(1):5–21

Chapter 2
Polyhydroxyalkanoates: The Future Bioplastics

Abstract Polyhydroxyalkanoates or PHAs are interesting biodegradable thermoplastics which are usually produced by bacteria intracellularly as an energy storage material under unfavourable growth condition. PHAs are attractive materials that can be developed as a bio-based commodity plastics. PHAs are also known as biocompatible polymers which can be used in various biomedical applications. This book chapter discusses about the production of PHAs by different types of bacteria using renewable resources.

Keywords PHAs production · Nutrient limitation · Gram negative bacteria · Gram positive bacteria · *Alcaligenes* · *Bacillus*

PHAs are valuable biopolymers with a broad range of applications in packaging, agriculture and medical field such as drug delivery and tissue engineering due to its biodegradability and thermo-processability. They have gained commercial attention because of their usage in packaging applications especially in plastic containers, bags, shampoo bottles and paper coatings (Reddy et al. 2003). PHAs also find uses in food packaging because of their interesting properties of resistance to moisture and impermeability to oxygen (Ebnesajjad 2012). Apart from their uses in packaging, PHAs have been considered as promising candidates for diverse medical applications because of its numerous advantages such as quick biological degradation, ease of processing and compatibility with animal tissue as compared to chemically produced polymers (Phillip et al. 2007). Moreover, the degradation product of PHB homopolymer is 3-hydroxybutyrate (3HB) which is a normal constituent of human blood (William and Martin 2002). As a result, PHAs are finding significant applications in biomedical field. Figure 2.1 summarizes various biomedical application of PHAs with important physiological properties.

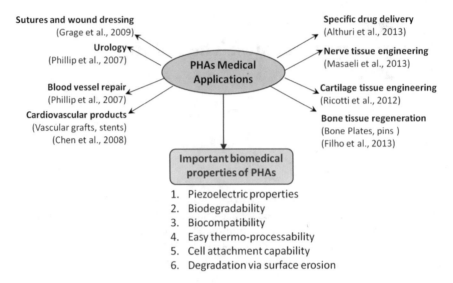

Fig. 2.1 Biomedical application of polyhydroxyalkanoates (William and Martin 2002; Peña et al. 2014)

2.1 PHAs Producing Microorganisms

Different types of gram-positive and gram-negative bacteria such as *Alcaligenes* sp., *Pseudomonas* sp., *Bacillus* sp. and *Methylomonas* sp. possess the unique capability to accumulate PHAs inside the cells as energy reserve materials in the form of inclusions. The biosynthesis of PHAs requires nutrient limitation of either nitrogen or phosphorous under excess carbon source availability. *Cupriavidus necator*, *Pseudomonas oleovorans* and *Methylobacterium organophilum* fall into this category (Akaraonye et al. 2010). However, some bacteria such as *Alcaligenes latus* and recombinant *E. coli* strain are capable of synthesizing PHAs during growth phase of the cultivation and do not essentially require nutrient limitation for PHA synthesis.

2.1.1 PHAs Production Using Gram-Negative Bacteria

2.1.1.1 PHAs Production Using *Cupriavidus necator* (Earlier Known as *Alcaligenes eutropha*)

Cupriavidus necator has been widely investigated for PHAs production because of its easy cultivability, well understood biochemistry and physiology and capacity to accumulate higher amounts of PHAs from renewable carbon sources such as glucose, fructose, and glycerol (Cavalheiro et al. 2009; Gahlawat and Soni 2017). It can accumulate PHB up to 80% of DCW when grown on inexpensive renewable resources such as olive oil, palm oil and waste frying oil (Loo et al. 2005). *C. necator* can also utilize specialized carbon sources like γ-butyrolactone and 1,4-butanediol and give rise to 4-hydroxybutyrate (4HB) monomers along with 3HB (Vigneswari et al. 2010). Moreover, cultivation on even number alkanoic acids and odd number alkanoic acid results in production of PHB homopolymer and copolymers of 3HB and 3-hydroxyvalerate (3HV), respectively (Khanna and Srivastava 2007). In 1970, Imperial Chemical Industries Ltd. (ICI/Zeneca BioProducts, Bellingham, UK) employed *R. eutropha* strain for the poly(3HB-co-3HV) production using propionate and glucose, and marketed it under the trade name of BIOPOL™.

Fed-batch and continuous cultivation of *C. necator* in bioreactor with valeric acid and sodium propionate feeding respectively demonstrated accumulation of PHA containing 3HV units (Khanna and Srivastava 2007; Yu et al. 2005). Cultivation in continuous stirred tank reactor helped in studying the cell growth kinetics and dynamic responses of poly(3HB-co-3HV) production (Yu et al. 2005). The terpolymers of poly(3HB-co-3HV-co-3HHx) have also been successfully synthesized from the mixture of crude palm oil and 3HV precursors using a mutant strain of *C. necator* PHB-4 transformant containing the PHA synthase gene (*phaC*) of locally isolated *Chromobacterium* sp. USM2 (Bhubalan et al. 2010). This group successfully produced a highly flexible material which demonstrated a high resemblance to low-density polyethylene by the incorporating both 3HV and 3-hydroxyhexanoate (3HHx) monomers in the PHA. *C. necator* DSM 545 can grow easily on by-products from biodiesel industry for microbial PHAs production (García et al. 2013). Utilization of crude biodiesel by-products as the main raw materials for PHAs production could improve the sustainability and viability of first generation fuel biorefinery.

2.1.1.2 PHAs Production Using *Alcaligenes latus*

A. latus produces PHAs during growth phase of cultivation at high rates, and can accumulate PHB up to 70–80% of CDW during cultivation (Gahlawat et al. 2012; Gahlawat and Srivastava 2012, 2013). It has also been indicated that PHB accumulation in *A. latus* can be increased further up to 88% of CDW by implementing nutrient limitation. Till date, the highest PHB productivity of 5.13 g/L h has been reported with *A. latus* DSM 1123 (Wang and Lee 1997), whose PHB synthesis rates were two to three times greater than the other bacteria. *A. latus* is able to grow on inexpensive, renewable resources rich in sucrose such as molasses and sweet sorghum juice for PHAs production (Tanamool et al. 2009; Zafar et al. 2012b).

A. latus ATCC29714 can synthesize copolymers of 3HB and 3HV using carbon sources of sucrose and valeric acid or propionic acid. A significantly high poly (3HB-co-3HV) content of 58% (w/w) and an overall 3HV content of 11 mol% was obtained using *A. latus* (Ramsay et al. 1990). It has been discovered that copolymers of 3HB and 4HB could be synthesized from mixed carbon substrate of sucrose and γ-butyrolactone using *A. latus* (Hiramitsu et al. 1993). The transformation yield of γ-butyrolactone into 4HB unit of copolymer by *A. latus* was observed to be approximately 60%. This report indicated that *A. latus* could be an excellent microbial system for the production of poly (3HB-co-4HB) on industrial level. Zafar et al. (2012a) demonstrated that *A. lata* MTCC 2311 can synthesize copolymers from cane molasses with enhanced P (3HB-co-3HV) accumulation. This study reported accumulation of 7.2 g/L concentration of poly (3HB-co-3HV) and 21 mol% of 3HV using cane molasses and propionic acid.

2.1.1.3 PHAs Production Using Other Gram-Negative Bacteria

Pseudomonas is a versatile species as it can utilize a broad range of substrates and the substrate specificity of its PHA synthase enzyme is very low which ultimately results in the diverse array of mcl-PHA monomers. *Pseudomonas* sp. can utilize variety of carbon substrates such as glucose, gluconate, hexane, heptane and octane etc. for PHAs production (Durner et al. 2001). However, carbon substrates such as glucose and gluconate are comparatively inexpensive and less toxic as opposed to alkanes. Cultivation of *P. putida* KT2442 on glucose resulted in accumulation of mcl-PHA containing mainly 3-hydroxydecanoate (3HD) and minor constituents of 3HHx, 3HO and 3-hydroxydodecanoate, (3HDD) (Kim et al. 2007). Some strains, for example *P. stutzeri, P. oleovorans* NRRL B-778 and *Pseudomonas* sp. DSY-82 not only accumulate mcl-PHAs but also copolymers containing both scl and mcl monomeric units, e.g. P(3HB-co-3HHx). *Pseudomonas stutzeri* 1317 demonstrated the ability to accumulate different types of PHAs when cultivated on glucose and mcl-fatty acids such as hexanoic acid (Guo-Qiang et al. 2001). Addition of different types of fatty acids as substrates helped in adjusting the monomer composition in PHAs such as addition of heptanoic acid during the growth on glucose and octanoic acid stimulated the accumulation of 3-hydroxyheptanoate (3HHp) in PHA. Thus *Pseudomonas* could be a better choice for production of copolymers with different monomer composition just by altering the different additional substrates (or fatty acids) in media.

Methylotrophs can also be employed for the production of PHAs as they grow very easily on the cheap carbon source, methanol. But they lack the ability to accumulate high amount of PHAs, thereby making the whole recovery process difficult. Therefore production of PHAs by methylotrophs can only be treasured if the productivity is increased by at least two-fold. In one study, *Methylobacterium* sp. GW2 isolated from groundwater was observed to be capable of synthesizing significant amount of PHA from methanol (Yezza et al. 2006). This GW2 strain was also able to synthesize copolymer poly(3HB-co-3HV) with a 3HB to 3HV ratio of 1:2 when valeric acid and methanol were used as carbon and energy sources. Moreover, molecular weight of copolymer poly(3HB-co-3HV) was higher than any other reported poly(3HB-co-3HV) copolymer from all methylotrophs investigated so far. In another report, *M. organophilum* UOCZ-2 strain, isolated from wastewater treatment plant, was capable of synthesizing substantial amounts of PHB using methane (Zúniga et al. 2011). Effect of nutrient limitation on culture was also studied in two-phase partitioning bioreactor (TPPB) to favor PHB production. This was the first ever report of methane degradation by *M. organophilum* in bioreactor with the simultaneous accumulation of PHB. In a recent study, the same isolate was utilized for the production of PHAs using methane and citrate or propionate as co-substrates (Zúniga et al. 2013). PHA synthesized with a mixture of methane and citrate comprised of six different monomer units namely 3HB, 3HV, 3-hydroxyoctanoate (3HO), 4HV, 4HO and 4HHp, demonstrating the great versatility of this PHA producing isolate. However, methylotrophs lack the ability to accumulate high amount of PHAs, thereby making the whole recovery process difficult and uneconomic. Production of PHAs by methylotrophs can only be beneficial if the overall productivity is increased by at least two-fold.

2.1.2 PHAs Production Using Gram-Positive Bacteria

2.1.2.1 PHAs Production Using *Bacillus* spp.

Earlier gram negative bacteria were the only producers of PHAs at commercial level but these bacteria contain lipo-polysaccharide (LPS) layer in their cell wall which co-purify with PHAs and have antigenic properties (due to presence of endotoxin) resulting in immunogenic response (Liu et al. 2014). In contrast, gram positive bacteria do not have LPS layer and hence can be used for biopolymer (PHAs) production particularly for biomedical applications. Among gram-positive bacteria, *Bacillus* spp. have gained commercial attention as PHAs producers as these bacteria can shift from PHB to copolymer production under different cultivations conditions, which mainly involves variations in raw substrate materials (Kumar et al. 2013). PHB yields and content from different *Bacillus* spp. are highly unpredictable and variable. PHB amount inside the cells varies from 2 to 83% of CDW. The maximum PHB accumulations by *Bacillus* spp. cultures on refined sugars, alcohols, volatile fatty acids (VFAs) and waste stream are approximately 83%, 64%, 80%, and 72% of CDW, respectively.

Among various refined sugars, glucose is the preferential substrate for *B. megaterium* for PHB production (López et al. 2012; Naranjo et al. 2013), followed by fructose for some other *Bacillus* spp. (Thirumala et al. 2010). López et al. (2012) characterized a newly isolated *B. megaterium* strain on the basis of 16S rRNA gene sequence and established its ability for PHB production by using different bioreactor configurations. The isolated strain accumulated PHB up to 59% of its CDW during cultivation on glucose as carbon source. Although *Bacillus* strains exhibited the sporulation phenomenon, yet the results demonstrated that the newly isolated strain can accumulate high amounts of PHB (López et al. 2012). Naranjo et al. (2013) also compared PHB production by *B. megaterium* using glucose and glycerol as substrates and obtained identical results. This report indicated the feasibility of glycerol as economic versatile carbon substrate for the PHB production at commercial scale.

In yet another report on *B. mycoides* cultivation, Narayanan and Ramana (2012) investigated the effect of glucose/peptone ratio on *B. mycoides* DFC1 with respect to biomass yield, PHB production and sporulation. Statistical optimization of process parameters showed 1.82-fold improvement in PHB concentration (from 1.83 to 3.32 g/L), amounting to PHB content of 76.32% of CDW. This report demonstrated that optimized process parameters encouraged maximum PHB accumulation and negligible sporulation in *B. mycoides*. *B. cereus* SPV strain can synthesize poly(3HB-co-3HV) using odd chain VFAs such as propionate, heptanoate, and nonanoate (Valappil et al. 2007). VFAs are excellent substrates for PHAs production, as compared to carbohydrates. The PHA yield are higher in case of long chain alkanoates such as nonanoate. The mole percent content (48 mol%) of 3HV monomer units was also highest when nonanoate was used as the substrate. However, the use of hexanoate and decanoate resulted in the formation of PHB. *B. cereus* SPV can utilize variety of fatty acids which indicates that it possess the capability of shifting the fatty acid metabolic cycle towards the PHA biosynthesis.

2.1.2.2   PHAs Production Using Other Gram Positive Bacteria

Among other gram-positive bacteria, *Streptomyces* sp. (Verma et al. 2002), *Corynebacterium* sp. (Alvarez et al. 2000), and *Staphylococcus* sp. (Wong et al. 2000) are also capable of producing PHAs. Verma et al. (2002) presented a correlation between PHB utilization and antibiotics (c-actinorhodin) synthesis in *Streptomyces coelicolor* M145 and suggested that PHB might function as a energy reserve compound for antibiotic synthesis. Thus it can be concluded that PHB, a primary metabolite, may serve as a key precursor for the synthesis of secondary metabolites during stationary phase of fermentation. More importantly, *Rhodococcus*, *Corynebacterium* and *Nocardia* are the only gram-positive bacteria which can synthesize copolymer, P (3HB-co-3HV) using simple carbon substrate, glucose without any precursors (Alvarez et al. 2000). This information could help in significant reduction of production cost of the copolymer. PHB has also been isolated from different species of *Staphylococci* (Wong et al. 2000). However, the exact role of PHA in these bacteria is unknown. Nevertheless, *Bacillus* genus appears to be an excellent candidate as opposed to other gram-positive members, for the PHAs production particularly for biomedical applications. This could be because of several reasons such as requirement of less stringent fermentation conditions, its capability to add both scl and mcl monomer units into PHAs polymer and better polymer yield.

References

Akaraonye E, Keshavarz T, Roy I (2010) Production of polyhydroxyalkanoates: the future green materials of choice. J Chem Technol Biotechnol 85(6):732–743

Alvarez HM, Kalscheuer R, Steinbüchel A (2000) Accumulation and mobilization of storage lipids by *Rhodococcus opacus* PD630 and *Rhodococcus ruber* NCIMB 40126. Appl Microbiol Biotechnol 54:218–223

Althuri A et al (2013) Microbial synthesis of poly-3-hydroxybutyrate and its application as targeted drug delivery vehicle. Bioresour Technol 145:290–296

Bhubalan K, Rathi DN, Abe H, Iwata T, Sudesh K (2010) Improved synthesis of P(3HB-co-3HV-co-3HHx) terpolymers by mutant *Cupriavidus necator* using the PHA synthase gene of *Chromobacterium* sp. USM2 with high affinity towards 3HV. Polym Degrad Stab 30:1–7

Cavalheiro JMBT, de Almeida MCMD, Grandfils C, da Fonseca MMR (2009) Poly (3-hydroxybutyrate) production by *Cupriavidus necator* using waste glycerol. Process Biochem 44:509–515

Chen QZ et al (2008) Biomaterials in cardiac tissue engineering: ten years of research survey. Mater Sc Eng R: Rep 59(1):1–37

Durner R, Zinn M, Witholt B, Egli T (2001) Accumulation of poly [(R)-3-hydroxyalkanoates] in *Pseudomonas oleovorans* during growth in batch and chemostat culture with different carbon sources. Biotechnol Bioeng 72(3):278–288

Ebnesajjad S (2012) Plastic films in food packaging: materials, technology and applications. Elsevier William Andrew Publishers, Oxford

Filho LX, Olyveira GM, Basmaji P, Costa LMM (2013) Novel electrospun nanotholits/PHB scaffolds for bone tissue regeneration. J Nanosc Nanotechnol 13(7):4715–4719

Gahlawat G, Soni SK (2017) Valorization of waste glycerol for the production of poly (3-hydroxybutyrate) and poly (3-hydroxybutyrate-co-3-hydroxyvalerate) copolymer by *Cupriavidus necator* and extraction in a sustainable manner. Bioresour Technol 243:492–501

Gahlawat G, Srivastava AK (2012) Estimation of fundamental kinetic parameters of polyhydroxybutyrate fermentation process of *Azohydromonas australica* using statistical approach of media optimization. Appl Biochem Biotechnol 168(5):1051–1064

Gahlawat G, Srivastava AK (2013) Development of a mathematical model for the growth associated polyhydroxybutyrate fermentation by *Azohydromonas australica* and its use for the design of fed-batch cultivation strategies. Bioresour Technol 137:98–105

Gahlawat G, Sengupta S, Srivastava AK (2012) Enhanced production of poly (3-hydroxybutyrate) in a novel airlift reactor with in situ cell retention using *Azohydromonas australica*. J Ind Microbiol Biotechnol 39(9):1377–1384

García IL et al (2013) Evaluation of by-products from the biodiesel industry as fermentation feedstock for poly(3-hydroxybutyrate-co-3-hydroxyvalerate) production by *Cupriavidus necator*. Bioresour Technol 130:16–22

Grage K et al (2009) Bacterial polyhydroxyalkanoate granules: biogenesis, structure, and potential use as nano-/micro-beads in biotechnological and biomedical applications. Biomacromolecules 10(4):660–669

Guo-Qiang C, Jun X, Qiong W, Zengming Z, Kwok-Ping H (2001) Synthesis of copolyesters consisting of medium-chain-length β-hydroxyalkanoates by *Pseudomonas stutzeri* 1317. React Funct Polym 48:107–112

Hiramitsu M, Koyama N, Doi Y (1993) Production of poly (3-hydroxybutyrate-co-4-hydroxybutyrate) by *Alcaligenes latus*. Biotechnol Lett 15:461–464

Khanna S, Srivastava AK (2007) Production of poly (3-hydroxybutyric-co-3-hydroxyvaleric acid) having a high hydroxyvalerate content with valeric acid feeding. J Ind Microbiol Biotechnol 34:457–461

Kim DY, Kim HW, Chung MG, Rhee YH (2007) Biosynthesis, modification, and biodegradation of bacterial medium-chain-length polyhydroxyalkanoates. J Microbiol 45(2):87–97

Kumar P, Patel SKS, Lee JK, Kalia VC (2013) Extending the limits of *Bacillus* for novel biotech-nological applications. Biotechnol Adv 31:1543–1561

Liu Y, Huang S, Zhang Y, Fuqian Xu (2014) Isolation and characterization of a thermophilic *Bacillus shackletonii* K5 from a biotrickling filter for the production of polyhydroxybutyrate. J Environ Sci 26:1453–1456

Loo CY, Lee WH, Tsuge T, Doi Y, Sudesh K (2005) Biosynthesis and characterization of poly (3-hydroxybutyrate-co-3-hydroxyhexanoate) from palm oil products in a *Wautersia eutropha* mutant. Biotechnol Lett 27:1405–1410

López JA et al (2012) Biosynthesis of PHB from a new isolated *Bacillus megaterium* strain: outlook on future developments with endospore forming bacteria. Biotechnol Bioprocess Eng 17:250–258

Masaeli E et al (2013) Fabrication, characterization and cellular compatibility of poly (hydrox-yalkanoate) composite nanofibrous scaffolds for nerve tissue engineering. PloS one 8(2):1–13

Naranjo JM, Posada JA, Higuita JC, Cardona CA (2013) Valorization of glycerol through the pro-duction of biopolymers: the PHB case using *Bacillus megaterium*. Bioresour Technol 133:38–44

Narayanan A, Ramana KV (2012) Polyhydroxybutyrate production in *Bacillus mycoides* DFC1 using response surface optimization for physico-chemical process parameters. 3 Biotech 2:287–96

Peña C et al (2014) Biotechnological strategies to improve production of microbial poly-(3-hydroxybutyrate): a review of recent research work. Microb Biotechnol 7(4):278–293

Philip S, Keshavarz T, Roy I (2007) Polyhydroxyalkanoates: biodegradable polymers with a range of applications. J Chem Technol Biotechnol 82(3):233–247

Ramsay BA, Lomaliza K, Chavarie C, Dube B, Bataille P, Ramsay JA (1990) Production of poly-(β-hydroxybutyric-co-β-hydroxyvaleric) acids. Appl Environ Microbiol 56(7):2093–2098

Reddy CSK, Ghai R, Rashmi KVC (2003) Polyhydroxyalkanoates: an overview. Bioresour Technol 83:137–146

Ricotti L et al (2012) Proliferation and skeletal myotube formation capability of C2C12 and H9c2 cells on isotropic and anisotropic electrospun nanofibrous PHB scaffolds. Biomed Mater 7(3):1–11

Tanamool V, Danvirutai P, Thanonkeo P, Imai T, KaewKannetra P (2009) Production of poly-β-hydroxybutyric acid (PHB) from sweet sorghum juice by *Alcaligenes eutrophus* TISTR 1095 and *Alcaligenes latus* ATCC 29714 via batch fermentation. In: The 3rd international conference on fermentation technology for value added agricultural products (FerVAAP), pp 1–6

Thirumala M, Reddy SV, Mahmood SK (2010) Production and characterization of PHB from two novel strains of *Bacillus* spp. isolated from soil and activated sludge. J Ind Microbiol Biotechnol 37:271–278

Valappil SP et al (2007) Polyhydroxyalkanoate (PHA) biosynthesis from structurally unrelated carbon sources by a newly characterized *Bacillus* sp. J Biotechnol 127:475–487

Verma S, Bhatia Y, Valappil SP, Roy I (2002) A possible role of poly-3-hydroxybutyric acid in antibiotic production in *Streptomyces*. Arch Microbiol 179:66–69

Vigneswari S, Nik LA, Majid MIA, Amirul AA (2010) Improved production of poly (3-hydroxybutyrate-co-4-hydroxybutyrate) copolymer using a combination of 1,4-butanediol and γ-butyrolactone. World J Microbiol Biotechnol 26:743–746

Wang F, Lee SY (1997) Poly (3-hydroxybutyrate) production with high productivity and high polymer content by a fed-batch culture of *Alcaligenes latus* under nitrogen limitation. Appl Environ Microbiol 63:3703–3706

Williams SF, Martin DP (2002) Applications of PHAs in medicine and pharmacy. In: Doi Y, Stein-büchel A (eds) Biopolymers polyesters III—applications, vol 4. Wiley-VCH, Weinhein, pp 1–38

Wong AL, Chua H, Yu PH (2000) Microbial production of polyhydroxyalkanoates by bacteria isolated from oil wastes. Appl Biochem Biotechnol 86:843–857

Yezza A, Fournier D, Halasz A, Hawari J (2006) Production of polyhydroxyalkanoates from methanol by new methylotrophic bacterium *Methylobacterium* sp. GW2. Appl Microbiol Biotech-nol 73:211–218

Yu ST, Lin CC, Too JR (2005) PHBV production by *Ralstonia eutropha* in a continuous stirred tank reactor. Process Biochem 40:2729–2734

Zafar M, Kumar S, Kumar S, Dhiman AK (2012a) Artificial intelligence based modeling and optimization of poly (3-hydroxybutyrate-co-3-hydroxyvalerate) production process by using *Azohydromonas lata* MTCC 2311 from cane molasses supplemented with volatile fatty acids: a genetic algorithm paradigm. Bioresour Technol 104:631–641

Zafar M, Kumar S, Kumar S, Dhiman AK (2012b) Modeling and optimization of poly (3hydroxybutyrate-co-3hydroxyvalerate) production from cane molasses by *Azohydromonas lata* MTCC 2311 in a stirred-tank reactor: effect of agitation and aeration regimes. J Ind Microbiol Biotechnol 39(7):987–1001

Zúniga C, Morales M, Borgne SL, Revah S (2011) Production of poly-β-hydroxybutyrate (PHB) by *Methylobacterium organophilum* isolated from a methanotrophic consortium in a two-phase partition bioreactor. J Hazard Mater 190:876–882

Zúniga C, Morales M, Revah S (2013) Polyhydroxyalkanoates accumulation by *Methylobacterium organophilum* CZ-2 during methane degradation using citrate or propionate as cosubstrates. Bioresour Technol 129:686–689

Chapter 3
Challenges in PHAs Production at Mass Scale

Abstract Polyhydroxyalkanoates (PHAs) biopolymers provide a suitable alternative to the synthetic plastics because of their biodegradability, biocompatibility and environment friendly manufacturing processes. PHAs are promising candidate for bio-based plastics because their material properties are quite similar to petroleum-based plastics and can be produced from renewable resources. However, the high production cost limit their application at industrial level. This book chapter discusses about the challenges faced by the society for the commercialization of biodegradable PHAs.

Keywords Biodegradable polymers · High cost · Low concentration · Recovery · Economical production

It is important to mention that despite the considerable efforts on production of PHB and its copolymers, only few commercial scale plants have been developed in the past decades. The main obstacle in the wider application of PHAs is high cost of production. At present, few enterprises around the world produce PHAs both at small and pilot scale levels. For example, Biomer Inc. (Germany), P&G (USA), PHB Industrial (Brazil) and Tianan Biologic (China) produce PHAs on small-scale with production quantities ranging from 100 to 50,000 ton/year, which increases its production cost (Chanprateep 2010; Wang et al. 2014). On the contrary, polypropylene is produced in volumes more than 300,000 ton/year, with a major benefit in the final cost due to the economy of scale used. The largest number of the companies are located in China. The Tianan company produces P(3HB-co-3HV) with the trade name Enmat. It had an estimated capacity of 2000 ton/year in 2007, but this is supposed to increase to 50,000 ton/year by 2020 (Alves et al. 2017). The Lianyi Biotech company produces PHBH, i.e., the polymer form of P(3HB-co-3HHx) under the trade name Nodax at small scale (2000 ton/year). Kaneka (Osaka, Japan) produces P(3HB-co-3HHx), marketed as Kaneka PHBH, with a capacity in 2007 of 100 ton/year; the forecast for 2020 is 50,000 ton/year. Danimer Scientific (Bainbridge, GA) produces Meredian PHA, which is a non-specific polymer and production of this polymer is expected to reach 272,000 ton/year by 2020. In Ulm, Germany, Biomers produced 10 ton/year in 2007 under the name Biomer at both the pilot and research scales. In 2010, the capacity of the overall bioplastics market in the Europe was 10,000–50,000 ton.

© The Author(s), under exclusive license to Springer Nature Switzerland AG 2019
G. Gahlawat, *Polyhydroxyalkanoates Biopolymers*,
Biobased Polymers, https://doi.org/10.1007/978-3-030-33897-8_3

Table 3.1 Global production of polyhydroxyalkanoate (PHA) (Chanprateep 2010; Alves et al. 2017)

Company	Country	Trade name	PHA type	Capacity in 2010 (ton/year)	Capacity in 2020 (ton/year)
Biomer Inc.	Germany	Biomer	P(3HB) and P(3HB-co-3HV)	50	n/a
Mitsubhishi	Japan	Biogreen	P(3HB)	10,000	n/a
Tianan Biologic	China	Enmat	P(3HB-co-3HV)	10,000	50,000
Kaneka Co.	Japan	Kaneka PHBH	P(3HB-co-3HHx)	1000	50,000
PHB Industrial	Brazil	Biocycle	P(3HB)	50	10,000 (disabled in 2015)
Zenica	England	Biopol	(P3HB-co-3HV)	Disabled in 1996	n/a
Monsanto	Italy	Biopol	P(3HB) and P(3HB-co-3HV)	Disabled in 1996	n/a
Telles	United States	Mirel	P(3HB)	50,000	500,000
Tianjin GreenBio	China	Green Bio	P(3HB-co-4HB)	10,000	n/a
Lianyi Biotech	China	Nodax	P(3HB-co-3HHx)	2000	n/a
Procter & Gamble	United States	Nodax	P(3HB-co-3HHx)	Disabled in 2006	20,000–50,000
Danimer	United States	Meredian	PHA from P&G	n/a	272,000

n/a—Data not available

At the end of 2020, the bioplastics market in the Europe is going to increase to 2–5 million ton. In Canada, the Biomatera (Toronto) company produced the polymer P(3HB-co-3HV), marketed with the name Biomater, at both the research and pilot scales. Tianjin GreenBio produced 1000 ton/year of P(3HB-co-4HB) copolymer. Table 3.1 gives the overview of PHAs manufacturers with their trade names, capacity (in ton) and market price (in €/kg).

While there has been expansion, many industries have stopped production of PHAs (Alves et al. 2017). For example, Zenica, located in England, produced Biopol polymer P(3HB-co-3HV), but suspended production in 1996; Monsanto, which produced the Biopol brand (P(3HB)/PHV/PHA) in Italy, ceased production in 1998. Procter & Gamble (Cincinnati, Ohio) produced Nodax P copolymer (3HB)/HHX, but ended this activity in 2006. However, it passed its production technology to Kaneca and Meredian, now Danimer. The technology Biopol, which belonged to Zeneka,

Constituents of PHA production cost

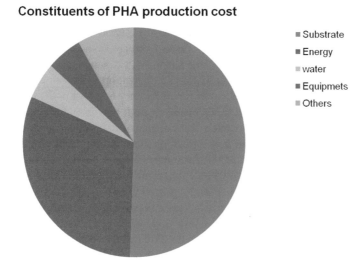

- Substrate
- Energy
- water
- Equipmets
- Others

Fig. 3.1 Important factors contributing in PHA production cost. Adopted from Wang and Chen (2017)

was transferred initially to Monsanto, which, in turn, transferred it to Metabolix (Cambridge, Massachusetts). In 2007, Metabolix and Archer Daniels Midland company (ADM; Chicago, Illinois) formed the Telles Company, which was dissolved in 2012. The Imperial Chemical Industries (London, England) produced the copolymer P(3HBHV) until 2008. Until recently, P(3HB) was produced in Brazil by the PHB SA company under the name BIOCYCLE. The company, located in Serrana, São Paulo, operated in partnership with the Institute of Technological Research, Copersucar, and the University of São Paulo. They used the microorganism *Ralstonia eutropha*, which was genetically modified for the growth on sucrose substrate. The company projected a capacity of 10,000 ton/year for 2020 but ceased production of P(3HB) in mid-2015.

PHA fermentation process cost mainly includes expense of substrate, energy, water, equipment and process complexity (Fig. 3.1). As per literature reports, the cost of substrate itself contribute to around 40–50% of the total PHAs production expense (Braunegg et al. 2004; Nath et al. 2008). Irrespective of the type of PHAs synthesized and amount of polymer produced, the issue of the production cost is always affecting the commercialization of PHAs. Thus, current research efforts are mainly geared towards making the entire production process economically feasible. The production cost can be reduced by taking several factors into consideration such as cell's ability to utilize inexpensive renewable carbon substrates for PHAs production, improved optimized fermentation processes, polymer synthesis rate, more efficient and economic cultivation strategies etc. (Pan et al. 2012). Recombinant strains are currently being exploited in order to obtain high PHA accumulation and high substrate conversion rates inside the host cell.

Other issues which need solution are the recovery and purification protocols that are currently being used. Various scientist have reported that around 50% of the expense is linked with the recovery of PHAs from cells (Jacquel et al. 2008; Pan et al. 2012; López-Abelairas et al. 2015). Therefore, extraction and purification cost should also be reduced for successful commercialization of PHAs. The final PHB concentration inside culture broths are generally very low within the range of few mg to 100 g/L. The low PHAs yields results in high downstream processing cost. On the contrary, polymer production from chemical industries can reach up to a level of 500 g/L. Therefore low cost extraction methods are desperately needed to significantly reduce the cost of PHAs. Moreover, product yield on substrate is very low in PHAs production processes (Penloglou et al. 2012). At industrial-scale, PHAs yield on the most generally used substrate is around 33% (g/g). On the other hand, in case of chemically synthesized plastics such as polyethylene and polystyrene etc. can be as high as over 90% or about to 100% (Domski et al. 2007). To obtain such high PHAs concentration, biomass concentration should be as high as greater than 150 g/L, besides polymer content should be higher than 90% of the CDW. Several investigators have developed different strategies to tackle this problem in order to achieve the benefits of chemical synthesis and obtain cost-effective PHAs production. These includes utilization of cheap renewable substrates (Castilho et al. 2009; Koller

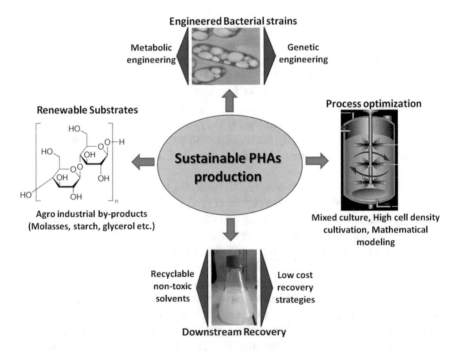

Fig. 3.2 Strategies for sustainable and cost-effective production of PHAs. Adopted from Gomez et al. (2012) with permission from IntechOpen

et al. 2008), efficient downstream recovery strategies (Fiorese et al. 2009; López-Abelairas et al. 2015), and various high cell density cultivation strategies (Ienczak et al. 2013; Cavalheiro et al. 2012; Kim et al. 2003). The main approaches that could be used for the inexpensive and sustainable production of PHAs are depicted in Fig. 3.2 and will be discussed in detail in next chapter.

References

Alves MI et al (2017) Poly (3-hydroxybutyrate)-P (3HB): review of production process technology. Ind Biotechnol 13(4):192–208

Braunegg G, Bona R, Koller M (2004) Sustainable polymer production. Polym Plast Technol Eng 43(6):1779–1793

Castilho LR, Mitchell DA, Freire DMG (2009) Production of polyhydroxyalkanoates (PHAs) from waste material and solid-state fermentation. Bioresour Technol 100:5996–6009

Cavalheiro JMBT et al (2012) Effect of cultivation parameters on the production of poly(3-hydroxybutyrate-co-4-hydroxybutyrate) and poly(3-hydroxybutyrate-4-hydroxybutyrate-3-hydroxyvalerate) by *Cupriavidus necator* using waste glycerol. Bioresour Technol 111:391–397

Chanprateep S (2010) Current trends in biodegradable polyhydroxyalkanoates. J Biosci Bioeng 110(6):621–632

Domski GJ, Rose JM, Coates GW, Bolig AD, Brookhart M (2007) Living alkene polymerization: new methods for the precision synthesis of polyolefins. Prog Polym Sci 32:30–92

Fiorese ML, Freitas F, Pais J, Ramos AM, de Aragão GM, Reis MA (2009) Recovery of polyhydroxybutyrate (PHB) from *Cupriavidus necator* biomass by solvent extraction with 1,2-propylene carbonate. Eng Life Sci 9(6):454–461

Gomez JG, Méndez BS, Nikel PI, Pettinari MJ, Prieto MA, Silva LF (2012) Making green polymers even greener: towards sustainable production of polyhydroxyalkanoates from agroindustrial by-products. (ed by Marian Petre) Adv Appl Biotechnol 41–62

Ienczak JL, Schmidell W, de Aragão GMF (2013) High-cell-density culture strategies for polyhydroxyalkanoate production: a review. J Ind Microbiol Biotechnol 40(3–4):275–286

Jacquel N, Lo C-W, Wei Y-H, Wu H-S, Wang SS (2008) Isolation and purification of bacterial poly(3-hydroxyalkanoates). Biochem Eng J 39(1):15–27

Kim M, Cho KS, Ryu HW, Lee EG, Chang YK (2003) Recovery of poly (3-hydroxybutyrate) from high cell density culture of *Ralstonia eutropha* by direct addition of sodium dodecyl sulfate. Biotechnol Lett 25:55–59

Koller M et al (2008) Polyhydroxyalkanoate production from whey by *Pseudomonas hydrogenovora*. Bioresour Technol 99:4854–4863

López-Abelairas M, García-Torreiro M, Lú-Chau T, Lema JM, Steinbüchel A (2015) Comparison of several methods for the separation of poly (3-hydroxybutyrate) from *Cupriavidus necator* H16 cultures. Biochem EngJ 93:250–259

Nath A, Dixit M, Bandiya A, Chavda S, Desai AJ (2008) Enhanced PHB production and scale up studies using cheese whey in fed batch culture of *Methylobacterium* sp. ZP24. Bioresour Technol 99(13):5749–5755

Pan W, Perrotta JA, Stipanovic AJ, Nomura CT, Nakas JP (2012) Production of polyhydroxyalkanoates by *Burkholderia cepacia* ATCC 17759 using a detoxified sugar maple hemicellulosic hydrolysate. J Ind Microbiol Biotechnol 39:459–469

Penloglou G, Chatzidoukas C, Kiparissides C (2012) Microbial production of polyhydroxybutyrate with tailor-made properties: an integrated modelling approach and experimental validation. Biotechnol Adv 30(1):329–337

Wang Y, Chen G-Q (2017) Polyhydroxyalkanoates: sustainability, production and industrialization. In: Tang C, Ryu CY (eds) Sustainable polymers from biomass. Wiley VCH, Weinheim, Germany, p 14

Wang Y, Yin J, Chen G-Q (2014) Polyhydroxyalkanoates, challenges and opportunities. Curr Opin Biotechnol 30:59–65

Chapter 4
Production Strategies for Commercialization of PHA

Abstract The most important criterion for large-scale production of polyhydroxyalkanoate (PHA) is sustainability in terms of supply and cost. The sustainable production of PHAs could be achieved by utilization of renewable, inexpensive carbon substrates and adopting efficient extraction processes. The operational cost of PHAs production process can be significantly minimized by using high yielding strains and various process optimization strategies. This chapter focuses on various strategies used in literature for cost-effective sustainable production of PHA.

Keywords Sustainability · Renewable substrates · Genetic engineering · Mathematical modelling · Downstream recovery

4.1 Use of Inexpensive and Renewable Substrates for PHA Production

PHAs production cost is significantly higher than the conventional plastics such as polystyrene and polyethylene. One of the most essential factors responsible for high production cost is the cost of major substrates. According to Aslan et al. (2016), around half of the production cost is due to the cost of pure substrates used during PHAs production. Production cost can be reduced by using renewable wastes materials and/or by-products from process industries as substrates. This mainly includes molasses Rathika et al. (2018), cheese whey (Pais et al. 2016), waste vegetable oils (Verlinden et al. 2011), crude glycerol (Cavalheiro et al. 2012), agricultural wastes (Amulya et al. 2015), palm oil (Rao et al. 2010), and waste water (Martinez et al. 2015). These renewable materials can act as promising substrate for economical production of PHA because of availability in huge quantities and high content of organic matter Verlinden et al. (2011). For example, around 1 gallon of crude glycerol is generated from every 10 gallon of biodiesel produced which is ultimately released as waste in environment. Thus, the crude glycerol can serve as interesting substrate for the production of PHAs which would not only help in the conversion of this waste by-product to a 'value added product' but will also make the biodiesel production process more sustainable. Bioconversion of glycerol to PHB has been investigated in detail by Cavalheiro et al. (2009) using *C. necator* DSM 545 which demonstrated accumulation of PHB concentration of 38.1 g/L with the waste glycerol. In another study, fed-batch cultivation of *C. necator* DSM 545 was used for the synthesis of copolymer poly (3HB-co-4HB) and terpolymers poly (3HB-co-4HB-co-3HV) using waste glycerol (Cavalheiro et al. 2012). The crude glycerol has also been successfully used by mixed-microbial culture for PHAs production in a feast-famine approach (Moralezo-Garate et al. 2013). The bacterial enrichment system has various benefits over pure microbial cultures such as allowing the use of low quality substrates and avoiding sterile operation thereby resulting in low overall production cost. Thus bacterial enrichment technique has become a recent trend these days for achieving good results in biopolymer production. Table 4.1 summarizes the list of various inexpensive raw materials used for PHAs production by different wild type and recombinant strains. The table also highlights the results of important kinetics parameters such as polymer content (% CDW) and productivity.

Apart from glycerol, whey and molasses are another potential inexpensive renewable substrates which have been studied largely for PHAs production. It is a major by-product of cheese and dairy industries and is considered as a waste material in many parts of the world. Nikel and co-workers (2006) demonstrated accumulation of PHAs by a recombinant *Escherichia Coli* K24K, containing *Azotobacter* sp. *phaC* biosynthesis genes using pH-stat fed-batch cultivation. This recombinant strain resulted in 70 g/L biomass concentration and accumulated huge amounts of 51 g/L PHB (72.9% DCW of PHB) when grown on concentrated whey solution. Obruca et al. (2011) investigated the ability of *Bacillus megaterium* CCM 2307 strain for PHB production using cheese whey wastes. Moreover, PHB production was increased around 50 fold after optimization of nutrient medium containing cheese whey (Table 4.1). However, the PHAs concentrations were relatively lower than that obtained in recombinant *E. Coli* cultivations. Molasses is a valuable by-product from sugar industry which can be used for PHAs production. Molasses contains various sugars (sucrose, fructose) and minerals such as iron, calcium, magnesium, potassium and vitamin B7 to promote the bacterial growth (Shasaltaneh et al. 2013). The usage of soy, beet and cane molasses for PHAs production is a promising alternative for their disposal (Solaiman et al. 2006). Recently, Rathika et al. (2018) elucidated that sugarcane molasses are remarkable substrate for the PHAs production by *Bacillus subtilis* RS1, and during batch cultivation a maximum biomass concentration of 9.5 g/L and PHA content of 70.5% of CDW were obtained with pre-treated cane molasses. *Alcaligenes latus* can also utilize beet and cane molasses for the production of PHAs at the rate comparable to *Alcaligenes eutrophus* (Zafar et al. 2012a, b, c).

Furthermore, the available literature reports suggest that the waste lipids can also serve as important substrate for PHAs production and still need to be investigated in detail. Lipids-rich wastes are generated in significant amounts through a variety of sources such as waste vegetable oils, spent palm oil and by-products from oil mills. It has been reported that *C. necator* ATCC 17699 has the ability to grow on spent palm oil obtained after frying and can synthesize poly (3HB-co-4HB) using 1,4-butanediol as precursor (Rao et al. 2010). After depyrogenation procedure, synthesized poly (3HB-co-4HB) blends showed biocompatibility with chick chorioallantoic membrane, which suggested that these blends could serve as absorbable biomaterial for

Table 4.1 Literature reports on PHAs production by various bacteria using inexpensive and renewable substrates

PHAs type	Microorganism used	Major substrates	Biomass (g/L)	PHAs (g/L)	PHA content (% CDW)	Process mode	References
P(3HB)	*Cupriavidus necator* DSM 545	Crude glycerol, rapeseed waste	19.5	10.9	55.6	Fed-batch	García et al. (2013)
	C. eutrophus B-10646	Carbon dioxide	50	42.5	85	Chemostat batch	Volova et al. (2013)
	C. necator DSM 428	Used cooking oil	10.4	3.8	37	Batch	Martino et al. (2014)
	Bacillus megaterium JK4h	Dry sea mix	0.611	0.342	55.97	Batch	Dhangdhariya et al. (2015)
	C. necator DSM7237	Crude glycerol	66.2	57	86.2	Fed-batch	Kachrimanidou et al. (2014)
	C. necator DSM 545	Olive mill wastewater	2.5	1.22	49	Fed-batch	Martinez et al. (2015)
	Recombinant *E. coli* XL1-Blue	Rice bran hydrolysate	2.98	2.68	90.1	Batch	Oh et al. (2015)
	Alcaligenes latus DSM 1123	Whey	1.51	1.28	84	Batch	Berwig et al. (2016)
	C. necator DSM 545	Crude glycerol	6.76	4.84	71	Batch	Gahlawat and Soni (2017)
	Pandoraea sp. MA03	Crude glycerol	9.1	5.73	63.7	Fed-batch	de Paula et al. (2017)
	B. cereus nm[*]	Food waste	0.76	0.029	3.9	Batch	Lam et al. (2017)

(continued)

Table 4.1 (continued)

PHAs type	Microorganism used	Major substrates	Biomass (g/L)	PHAs (g/L)	PHA content (% CDW)	Process mode	References
	C. necator DSM4058	Crude glycerol	28.9	24.75	85.6	Fed-batch	Salakkam and Webb (2018)
P(3HB-co-3HV)	A. lattus MTCC 2311	Cane molasses	10.25	6.25	61.0	Batch	Zafar et al. (2012b)
	Haloferax mediterranei DSM1411	Crude glycerol	20.2	15.2	75.4	Fed-batch	Harmann-Kraus et al. (2013)
	B. mycoides DFC1	Rice husk	1.1	0.38	34.5	Batch	Narayanan et al. (2014)
	H. mediterranei DSM 1411	Rice-based ethanol stillage	n/a	16.42	71	Batch	Bhattacharyya et al. (2014)
	H. mediterranei ATCC 33500	Cheese whey	7.54	7.92	53	Batch	Pais et al. (2016)
	Anaerobic consortia	Food waste	n/a	n/a	23.7	Fed-batch	Amulya et al. (2015)
	Hydrogenophaga palleronii NBRC102513	Synthetic wastewater	1.61	1.01	63	Batch	Reddy et al. (2016)
	H. mediterranei DSM 1411	Olive mill waste water	10.0	0.2	43	Batch	Alsafadi and Al-Mashaqbeh (2017)
	Halogeometricm borinquense E3	Cassava waste	3.4	1.52	44.7	Batch	Salgaonkar et al. (2019)
	Paracoccus sp. LL1	Waste cooking oil	3.24	1.0	30.9	Batch	Kumar and Kim (2019)

(continued)

Table 4.1 (continued)

PHAs type	Microorganism used	Major substrates	Biomass (g/L)	PHAs (g/L)	PHA content (% CDW)	Process mode	References
P(3HB-co-4HB)	*C. necator* ATCC17699	Spent palm oil	5.4	4.37	81	Batch	Rao et al. (2010)
	C. necator DSM 545	Crude glycerol	30.0	10.9	36.1	Fed-batch	Cavalheiro et al. (2012)
	Cupriavidus sp. USMAHM13	Glycerine pitch, 1,4 butanediol	6.4	3.07	48	Batch	Ramachandran and Amirul (2013)
	Methylocystis parvus OBBP	Methane, ω-hydroxybutyrate	n/a	n/a	50	Batch	Myung et al. (2017)
	Halomonas bluephagenesis TD01	Waste corn-steep liquor, gluconate	90	66.6	74	Fed-batch	Ye et al. (2018)
P(3HB-co-3HHx)	Recombinant *E. coli* ABC$_{Ah}$	Crude glycerol	4.0	0.6	14	Batch	Phithakrotchanakoon et al. (2015)
	Ralstonia eutropha Re2133	Coffee waste oil	0.93	0.64	69	Batch	Bhatia et al. (2018)
	R. eutropha Re2133	Food waste	1.1	0.57	52	Batch	Bhatia et al. (2019)

* nm means strain not mentioned

biomedical uses. In another study, waste frying oil was successfully converted to PHB by *C. necator* resulting in an overall PHB content of 38% DCW (Verlinden et al. 2011). Using waste frying oil from food industry could be both cost-effective and environmentally advantageous. *C. necator* H16 strain showed accumulation of poly (3HB-co-3HV) copolymer when cultivated on rapeseed oil along with propionic acid (Obruca et al. 2010). Plant oils like olive, palm, sunflower and corn oil have also been recognized as potential renewable feedstock for the biopolymer production (Ciesielski et al. 2015). Through life cycle analysis, plant oils were found to have an advantage over conventional carbon feedstock like sugars in terms of their cost effectiveness and their capability to offer high yield of PHA (Loo and Sudesh 2007). However, more investigations on the development of genetically engineered vegetable and plant oil utilizing strains and better cultivation strategies for utilization of plant oils are required to further increase the PHA accumulation with these resources.

Agricultural crop residues wastes represent a vast biomass and, thus renewable resource that could provide an invaluable feedstock for biopolymer production (Amulya et al. 2015). These wastes include sugarcane industry waste, wheat bran, rice bran, rice husk, food waste, paper-mill wastes, tomato cannery waste etc. Oh et al. (2015) demonstrated that rice bran hydrolyzate can be utilized as a suitable alternative carbon feedstock for PHAs production using *R. eutropha* NCIMB 11599 and recombinant *E. coli*. In this study, a significantly high PHB content of 97.2 and 90.1% of CDW was achieved with *R. eutropha* NCIMB 11599 and recombinant *E. coli* respectively using saccharified rice bran. This study revealed the rice bran can serve as a suitable renewable substrate for biomass-based production of PHB. In the past decade, continuous research efforts have been devoted globally to implement bio-refinery strategies for the conversion of cellulosic and lignocellulosic waste into sustainable feedstock (substrates) for the production of PHA polymers (Liu et al. 2012). But the major obstacle(s) in PHA production from lignocellulosic waste are the resistance of such materials to hydrolytic treatments to release the fermentable hexoses and pentoses and the presence of inhibitory molecules (e.g., phenolics from lignin) which significantly affect the microbial growth and PHB accumulation.

Various literature reports have summarized the possibility of using the other cheap substrates such as paper-mill waste water, starch and malt waste for the biopolymer production. Alsafadi and Al-Mashaqbeh et al. (2017) elucidated the synthesis of poly (3HB-co-3HV) copolymer by halophilic culture of *Haloferax mediterranei* using olive mill wastewater (OMW) in one stage cultivation without any additional step of dephenolization. It was reported that *H. mediterranei* can bear the inhibitory concentration of polyphenols in waste up to 25% of OMW. Haas et al. (2008) demonstrated that saccharified potato starch can be utilized as a renewable substrate for PHB production using *R. eutropha* NCIM 5149. In this study, a very high cell concentration of 179 g/L and PHB concentration of 94 g/L was achieved during high cell density fed-batch cultivation. It has been reported that *A. latus* DSM 1124 could utilize industrial food waste such as malt and soya waste from beer brewery plant and soya milk dairy respectively for the production of biodegradable PHAs (Yu et al. 1999).

Recently, there has been more focus on the utilization of industrial waste gases such as CO_2, and methane for the production of PHA. These carbon sources can avoid the problem associated with exploitation of food crops and dependency on plant-based carbon sources. Moreover, the production process will be environmentally friendly and cost-effective. It has been elucidated that CH_4 and CO_2 are the most important greenhouse gases till date. The level of these gases will keep on rising with time due to continuous increase in industrialization, urbanization and population growth. In earlier reports, PHAs production using gases as carbon source required more energy consumption, thus was more costly than the food crop based substrates. In Herrema and Kimmel (2012) demonstrated a unique strategy for PHA production by methanotrophic consortium using methane and CO_2 as substrates. This report showed that the concentration of copper in the medium plays a significant role in the PHAs production, as it helps in controlling the switchover between pMMO (particulate methane monooxygenase) and sMMO (soluble methane monooxygenase) production. By decreasing the concentration of copper in the medium, PHA can be synthesized preferably using pMMO. This strategy is advantageous as pMMO based PHA synthesis has higher rates as compared to sMMO. Moreover, they have adopted site directed mutagenesis for genetic engineering of the microorganisms to have limited sMMO expression and high expression of pMMO gene. A high intracellular concentration of 94.5 g/L PHA or 70% PHA of CDW was obtained. The main finding of this report was increased PHA production achieved through a feed forward strategy. The other key finding of this report was that the technique of gas-based mixing was used which played an important role in reducing the shear stress inside the vessel. Another drawback of using methane as substrate is its low solubility in media which restricts both mass transfer and availability to microbial cells for PHAs production. Zúñiga and coworkers (2011) have adopted two phase partitioning bioreactor (TPPB) strategy to sort out this problem. This report studied the PHA production from methane gas using methanotrophic bacteria isolated from a wastewater treatment plant. Silicone oil was used to increase the solubility of methane in the culture medium thereby resulting in increased PHA production. The specific productivity was observed to be as high as 1.83 mg(PHB)/mg(CH_4).h. Therefore, in coming years researchers will now focus more on improving the diffusion rates or solubility of the gases into culture medium. Table 4.1 summarizes the literature reports of PHAs production by different bacteria using various inexpensive and renewable substrates. Although the employment of low-cost waste materials for PHAs production is beneficial, the presence of certain inhibitory compounds and the changes in waste by-products composition every time generates variation in the final PHAs concentration, recovery yields and content (Du et al. 2012). Such type of variation from these fermentation processes could be avoided by performing prior PHA optimization strategies at lab-level before going for pilot scale testing.

4.2   Engineering of Superior PHA Producing Strains

Recent advances in metagenomics and molecular biology helps in construction of superior PHAs producing strains to cater specific technological demands for reduction of production cost of PHAs. An industrial strain for PHA production should have the following characteristics: non-pathogenic, clear genomic background, fragile cell wall, easy genetic manipulation, easy flocculation, fast growing in a defined medium, no toxin production and possibly cellulose utilizing (Wang et al. 2014). With the above indispensable properties, it is not so difficult to incorporate other essential features for economical production e.g. high PHA accumulation capacity (80–90% of CDW), growth on mixed substrate, high substrate to polymer yield and production of SCL and/or MCL PHA. For efficient recovery of PHAs, the genetically modified strain should have a fragile cell wall, large cell size and easy or inducible flocculation (Kourmentza et al. 2017). Figure 4.1 shows desirable properties of an ideal strain for PHA industrial production. Thus, efforts of PHA commercialization mainly focuses on engineering strains that exhibit high PHA accumulation from renewable feed-stock, consume less energy for PHA synthesis, efficient and easy PHA recovery strategies, and tailored synthesis of functional PHA structures for high end applications.

Recombinant *Escherichia coli* have been one of the most desirable microorganism for the polymer synthesis because of growth associated PHA production to higher intracellular levels. Usage of *E. coli* for the PHAs production has several other benefits such as fast growth, absence of depolymerase enzyme, ability to utilize versatile inexpensive carbon substrates like waste glycerol and molasses, and easy recovery of polymer from cells (Reddy et al. 2003; Leong et al. 2014). With the tremendous knowledge about the *E. coli*'s genome and metabolic events concerning PHA production, it is anticipated that *E. coli* will keep on playing key role in hastening

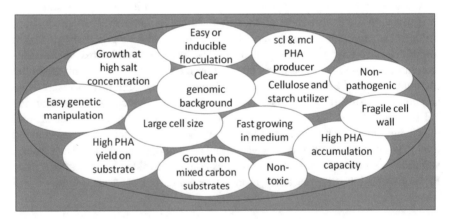

Fig. 4.1 Desirable properties for a PHA industrial production strain. Reproduced from Wang et al. (2014) with permission from Elsevier

the molecular analysis of PHA biosynthesis process and the industrial production of PHAs (Akaraonye et al. 2010). Therefore, *E. coli* could be commercially exploited using genetic engineering approaches to generate biopolymer with better physical and mechanical properties.

Researchers have reported that modification of genes responsible for oxygen uptake, quorum sensing, and PHA biosynthesis operons can improve PHA accumulation (Kourmentza et al. 2017). During high cell density cultivation, oxygen level has always been a concerning factor, therefore separating cell growth stage from an oxygen limitation induced PHA production phase will be an appropriate solution to increase the PHA accumulation along with high yield of PHA. The stage separation of cell growth from PHA accumulation can be achieved by inserting PHAs synthesis operons after anaerobic or microaerobic promoter (Wei et al. 2009). Wei et al. (2009) cloned nine anaerobic promoters upstream of PHB synthesis operon *phb*CAB (consist of *phbA, phbB, phbC* genes) from *R. eutropha* in recombinant *E. coli*. Amongst all the promoters, the one coding for alcohol dehydrogenase (PadhE) enzyme was observed to be highly efficient. Recombinant *E. coli* JM 109 containing PadhE promoter and *phb*CAB operon obtained 48% PHB production inside the cell in 48 h of cultivation whereas only 30% PHB accumulation was achieved under its original promoter. Li et al. (2016) cloned *phb*CAB operon from *R. eutropha* strain into *E. coli* for PHB pathway optimization. It was investigated that pathway optimization can be done by fine tuning expression of the three genes namely *phb*C (PHA synthase), *phb*A (3-ketothiolase), and *phb*B (acetoacetyl Co-A reductase). Ribosomal Binding Sites (RBS) libraries for all three genes were designed in a single-step reaction by an Oligo-Linker Mediated Assembly method (OLMA). Recombinant strains with desirable characteristics were screened and chosen on the basis of three different approaches such as visual selection, high-throughput screening and comprehensive analysis. This study proved that the semi rational and integrated approach of library designing and construction, and high throughput screening are an effective tool to improve PHB accumulation.

Genetic engineering approach has also been adopted for manipulating the morphology of bacteria for increasing the cell size for bioaccumulation of PHAs (Wang et al. 2014). Apart from PHAs inclusion bodies, numerous bacteria can synthesize granules of polyphosphates, elemental sulfur, glycogen, or proteins inside their cells that restricts cell space for PHA accumulation (Chen and Jiang 2018). Thus, a lot of research work have been done to increase the cell size of the PHA accumulating bacteria. For example, Jiang et al. (2015) have focussed on increasing cell intracellular space by deleting or weakening the expression of an actin-like protein, MreB in *E. coli* which ultimately showed enhancement in PHB accumulation by 100%. Moreover, engineering of PHAs-associated proteins demonstrated increase in PHA granule size, thus facilitating easy extraction and recovery (Pfeiffer and Jendrossek 2012). Similarly, Jiang et al. (2017) developed a temperature-dependent plasmid expression system for *mreB* gene (encodes for cytoskeleton protein) and *ftsZ* gene (encodes for cell division protein) in *Halomonas campaniensis* LS21. It was observed that *H. campaniensis* was able to grow normally in cell size at 30 °C in the presence of *mreB* and *ftsZ* genes, but cell size expansion became evident at 37 °C upon loss of *mreB* and *ftsZ* genes encoded by pTKmf plasmid. It was inferred that controlling cell size provide more space for PHA accumulation and gravity induced cell separation help in downstream recovery of the cells and product.

Recently, halophilic bacteria have been exploited for PHA production under continuous and unsterile cultivation conditions using synthetic biology approach (Yue et al. 2014; Chen et al. 2017). These characteristics enhance the competitiveness of PHA production at commercial level. Moreover, halophilic bacteria are easy to manipulate by genetic engineering approach, thus facilitating the design of a hyper-producing strain (Wang et al. 2014). In one study, both recombinant and wild type strains of *Halomonas campaniensis* LS21 were found to be capable of growing on kitchen wastes (mixed substrates) at 27 g/L NaCl concentration under 10 pH and 37 °C for 65 consecutive days (Yue et al. 2014). Recombinant *H. campaniensis* yielded approximately 70% PHB inside the cells in comparison to wild type wherein PHB accumulation was only 26% of CDW. Tao et al. (2017) have adopted an efficient approach of CRISPRi (Clustered regularly interspaced short palindromic repeats interference) for selective inhibition of genes in *Halomonas* species TD01. CRISPRi strategy was successfully used to inhibit expression of *ftsZ* gene encoding for cell division protein, ultimately resulting in an enlarged cell morphology for improvement in PHA accumulation. Thus, *Halomonas* sp. TD01 appears to be a promising strain for production of different PHAs types under open, unsterile and continuous cultivation conditions.

PHA biosynthesis generally competes with other metabolic cycles and intermediates, therefore it is essential to delete or weaken competing pathways so that more substrates are fluxed towards PHA biosynthesis pathway. Li et al. (2010) have demonstrated that metabolically engineered *E. coli* could synthesize P (3HB-*co*-4HB) copolymer using co-expression of succinate degrading genes from *Clostridium kluyveri* and PHB formation genes from *R. eutropha*. In this case, *E. coli* succinate semialdehyde dehydrogenase gene (*gabD*) were knocked out to block the pathway of succinate synthesis in order to channel the carbon flux to 4HB synthesis. This approach was helpful in production of P (3HB-co-4HB) using glucose only and reduced the dependency of microbe on much more expensive 4HB-precursors required for copolymers synthesis. Researchers have focussed on synthesis of functional PHA structures such as random and block copolymers to expand PHA diversity for different applications. Block copolymers have been studied extensively because of their resistance against polymer deterioration that affect polymer properties (Tripathi et al. 2012, 2013). This delay in degeneration of polymer results in better polymer performance and consistency in their properties. Tripathi et al. (2012) first time reported production of block copolymers of P(3HB)-b-P(3HHx), comprising of PHB block covalently bonded to poly-3-hydroxyhexanoate (PHHx) block, by recombinant strain *P. putida* KT2442 with β-oxidation pathway deleted from its genome. It was discovered that down regulation of β-oxidation pathway genes (*fadA* and *fadB*) in *P. putida* and *P. entomophila* can be helpful in tailoring PHA monomer composition using different fatty acid precursors for PHA synthesis (Tripathi et al. 2013).

Li et al. (2014) also constructed a β-oxidation deleted recombinant *P. entomophila* for the production of random copolymers of 3-hydroxydodecanoate (3HDD) and 3-hydroxy-9-decenoate (3H9D) by adjusting ratios of dodecanoic acid to 9-decenol precursors. This diversification in PHAs structures is achievable by manipulating either general biosynthesis pathways (acetoacetyl-CoA pathway, β-oxidation pathway, and/or in situ fatty acid synthesis) or through the specificity of PHA synthase enzyme. Thus, different approaches of genetic manipulations have been used in literature to improve PHA accumulation with high productivity and to synthesize diverse range of copolymers. Wang et al. (2017) engineered a β-oxidation pathway deletion mutant of *P. entomophila* strain for the channelling of all fatty acids towards synthesis of mcl-PHA homopolymers. The utilization of this engineered strain resulted in synthesis of whole range of mcl-PHA homopolymers ranging from C7 to C14, random copolymers of 3-hydroxyoctanoate (3HO) and 3-hydroxydodecanoate (3HDD) and block copolymers of P(3HO)-b-P(3HDD) using different ratios of fatty acids precursors. These synthesized mcl-PHAs are high value-added products and have much wider applications in healthcare sector than scl-PHAs. Figure 4.2 gives overview of various strategies adopted for manipulation of bacterial cells to improve PHAs production.

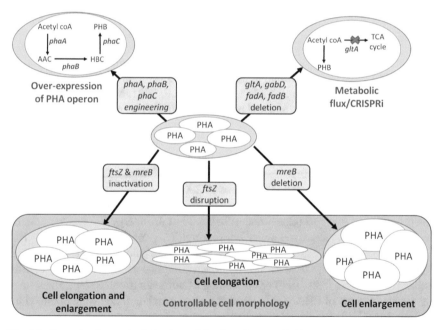

Fig. 4.2 Strategies for manipulating bacterial cells to improve PHAs production. Note *mreB* gene encodes for cytoskeleton protein; *ftsZ* gene encodes for cell division protein; *gabD* encodes for succinate semialdehyde dehydrogenase in succinate pathway; *fadA* and *fadB* encodes for thiolase and hydroxyacyl-CoA dehydrogenase enzymes in β-oxidation pathway; *gltA* encodes for citrate synthase in TCA cycle; *phaA* encodes for β-ketothiolase; *phaB* encodes for NADPH-dependent acetoacetyl-CoA reductase; *phaC* encodes for PHA synthase gene; AAC-Aceto-acetyl-CoA; HBC-3-hydroxybutyryl-CoA. Adopted from Chen and Jiang (2018) with permission from Elsevier

4.3 PHAs Production by Mixed Cultures

The large-scale production of PHA is generally carried out using pure culture or recombinant strains and pure substrates (Koller et al. 2016). Another interesting approach that could help in lowering down the PHA production cost is the employment of mixed culture strategy (Bugnicourt et al. 2014; Oliveira et al. 2016). This strategy utilizes open non-sterile mixed microbial consortium (MMC) and nature's principles of selection and competition, where microorganisms capable of accumulating PHA are selected on the basis of cultivation conditions applied on the consortia. Hence, the main aim is to manipulate the whole biological system, rather than particular organism, combining the technique of evolutionary engineering with the objectives of bioprocess technology (Johnson et al. 2009). The cost reduction is mainly achieved by performing cultivations under non-sterile energy saving conditions, and utilizing the high adaptation capability of MMC to grow on inexpensive complex feedstock such as industrial and domestic wastes. Fermentation processes for PHA accumulation in MMC are generally executed in three stages. In the first stage, substrates are initially converted into volatile fatty acids (VFAs) in continuous stirred tank reactor (CSTR). The mixed cultures utilize VFAs as the precursors for PHA synthesis, whereas the pure cultures generally use glucose as a substrate for PHA synthesis (Van Loosdrecht et al. 1997). The main reason behind this is that using carbohydrates in MMC and other substrates such as glycerol leads to formation of glycogen granules in addition to PHAs (Dircks et al. 2001; Moralejo-Gárate et al. 2011). In the second stage, sequential batch reactor (SBR) is employed to selectively enrich particular microbial population having high PHAs accumulation potential by implementing transitory conditions. In third stage, the biomass obtained from second stage was imposed to conditions favouring PHAs production, and after that cellular biomass is separated for PHA extraction and recovery when culture have achieved maximum PHA accumulation (Johnson et al. 2009).

The mixed culture systems are gaining attention due to their high PHA accumulation capability at low cost (Chanprateep 2010). PHA are energy storage compounds which are usually synthesized under limiting nutrients conditions, therefore cyclic feast-famine approach can be adopted as a selective technique for PHA accumulation. This type of cyclic feast-famine approach comprise of continually alternating feeding of nutrients (feast step) or no feeding of nutrients (famine step). Encouraging results in terms of PHA yields, content and accumulation rate were obtained by growing MMC either on simple media like acetate (Serafim et al. 2004), or on complex media obtained from industries wastes such as olive mill wastewater (Ntaikou et al. 2014; Kourmentza et al. 2015), cheese whey (Duque et al. 2014), sugar cane molasses and food wastes (Amulya et al. 2015) etc. These complex substrates contain high amounts of sugars, vitamins, and/or fatty acids which are either used for MMC-mediated PHA production, or pre-fermented towards volatile fatty acids (VFAs). Valentino et al. (2017) have reviewed and discussed various studies on mixed cultures for PHA production using industrial waste streams. Moita and Lemos (2012) elucidated successful synthesis of copolymer containing 70% of 3HB monomers and

30% of 3HV monomers by β-proteobacteria using bio-oil derived after pyrolysis of chicken-beds by MMC under feast-famine conditions. Fradinho et al. (2013) discovered an interesting method of PHA synthesis under feast and famine conditions wherein acetate was used as carbon source and consortium of photosynthetic algae along with bacteria was used as oxygen source. This opens up a new possibility of using sunlight driven PHA production without using any external oxygen source to lower the cost of production. Recently, enrichment of halophilic bacterial population for PHA accumulation has been performed under aerobic dynamic feed conditions using different carbon substrates, which yielded an overall PHA contents of around 66 and 62% on acetate and glucose, respectively, thus revealing the efficiency of MMC in PHA accumulation (Cui et al. 2016). PHA synthesis by halophiles can be accomplished by using seawater instead of pure water or using highly saline wastewater released by various food processing industries. Moreover, distilled water can be used for the lysis of halophilic bacterial cells thus helping in reduction of downstream processing costs (Kourmentza et al. 2017).

Furthermore, relevant research efforts have recently been started (2010 onwards), and few studies have reported PHA production by MMCs at pilot-scale level, while no industrial level PHA production by MMC has been reported till now. A standard characteristic in all pilot-level studies is that effluent streams are always subjected to fermentation before PHA accumulation. The pilot-scale studies on PHA production mainly focus on combining PHA synthesis with already available approaches on industrial waste treatment plants so as to lower the cost of production. The first pilot-level study on PHA production by MMC was carried out using fermented milk and ice-cream industry wastes which resulted in a maximum PHA content of 43% of CDW and PHA yield of 0.25 g/g COD (Chakravarty et al. 2010). Tamis et al. (2014) evaluated the potential of wastewater from candy bar factory for PHA accumulation. The activated sludge was used for enrichment into PHA producing microorganisms using feast-famine conditions, and the total PHA content at the end of cultivation was reported to be 70–76%, which is the highest reported till now at pilot-scale level. In another study, the denitrifying microbial culture was used for PHA accumulation by adopting feast-famine regime of anorexia and aerobic condition, respectively, and thus the strategy involved both nitrification and denitrification process for PHAs accumulation (Bengtsson et al. 2017). Table 4.2 depicts the details of literature reports on pilot-scale studies for PHA production by MMC. The performance of pilot-scale level cannot be compared with lab-scale level but they give crucial information on PHA synthesis under different compositions of raw material and leads to accumulation of significantly high amounts of PHA. The variation in raw material composition along with limitations in oxygen mass transfer at higher scale are the main reasons for low PHA yields and productivities at pilot-scale level in comparison to lab-scale studies.

Table 4.2 Main characteristics of PHA production by mixed microbial consortium at pilot-scale

Pilot plant location	Feedstock	Origin of MMC and enrichment strategy	$Y_{P/S}$ (g/g)*	PHA % (%mol HB; %mol HV)	mgPHA/gX/h	References
Nagpur, India	Pre-fermented milk and ice cream processing wastewater	Activated sludge	0.425*	39–43	–	Chakravarty et al. (2010)
Lucun WWTP in Wuxi, China	Hydrolyzed and acidified raw excess sludge	Activated sludge/synthetic mixture of VFA, ADF feast famine with carbon limitation and inhibitor of nitrification	0.044–0.29*	–	2.06–39.31	Jia et al. (2014)
Eslöv, Sweden	Beet process water, 38% in VFA	PHA producing MMC from pre-fermented effluent of Procordia Foods	–	60(85:15 HB:HV)	–	Anterrieu et al. (2014)
Brussels North WWTP (Aquiris, Belgium)	Pre-hydrolyzed and fermented WWTP sludge	Sludge fed with municipal waste water under aerobic feast famine	0.25–0.38	27–38(66–74:26–34 HB:HV)	100–140	Morgan-Sagastume et al. (2015)

(continued)

Table 4.2 (continued)

Pilot plant location	Feedstock	Origin of MMC and enrichment strategy	$Y_{P/S}$ (g/g)*	PHA % (%mol HB; %mol HV)	mgPHA/gX/h	References
Leeuwarden WWTP, Friesland, Netherlands	Fermented residuals from green-house tomato production	Sludge fed with municipal waste water under anoxic feast/aerobic famine	0.30–0.39	34–42(51–58: 42:49HB:HV)	28–35	Bengtsson et al. (2017)
Mars company, Veghel, Netherlands	Fermented wastewater from a candy bar factory	Activated sludge from a WWTP fed with the fermented wastewater under aerobic feast/famine with inhibitor of nitrification	0.30	70–76(84:16 HB:HV)	–	Tamis et al. (2014)
Milan, Italy	Organic fraction of municipal solid waste (OFMSW)	Organic acid-rich percolate from OFMSW under ADF	0.44	(53:47 HB:HV)	–	Colombo et al. (2017)

*Yield calculated on a COD basis by using the coefficients: for HB: 1.67 g COD PHA/g PHA and for HV: 1.92 g COD PHA/g PHA. *VFA* volatile fatty acids, *WWTP* waste water treatment plant; *ADF* aerobic dynamic feeding

4.4   High Cell Density Cultivation Approaches

The batch fermentation seems to be an easy cultivation strategy for the production of industrially important metabolites. But it has major disadvantages like substrate limitation problem towards the end of cultivation and low cell yields during the cultivation which restricts the usage of batch cultures at industrial level. Therefore, there is need to develop improved cultivation strategies which could help in the cultivation under non-limiting and non-inhibitory substrate concentrations and subsequently enhance the total yield and productivity of the PHB accumulation during the cultivation. Fermentation processes with high cell densities are highly desirable as they favour enhanced PHAs accumulation, especially with significant reduction in the culture volume, reduction in production costs, equipment cost, and in waste water production. High cell density cultivations involve several important aspects such as media optimization, reactor types and operating conditions, types of fermentation modes, feeding and modelling strategies. Continuous and fed-batch cultivations are the two important operation approaches for high cell density cultivation of microorganisms. These cultivations yield high PHAs concentrations and productivity. Table 4.3 summarizes the results of different high cell density cultivation strategies for PHAs production.

Cavalheiro et al. (2012) adopted a high cell density culture fed-batch cultivation strategy to attain high PHAs concentrations and productivity. The main carbon source for culture growth was waste glycerol derived from the production of biodiesel from refined vegetable oils. The fed-batch strategy yielded 45 g/L total biomass and 16.7 g/L PHAs concentrations and final PHB productivity of 0.25 g/L h. This is the first report that has achieved high cell densities by fed-batch using by-product glycerol. In another report, Ienczak et al. (2011) have also demonstrated that maintenance of carbon concentration at optimum level during the cultivation is extremely important. The authors established that pulse addition of carbon should not be attempted after carbon exhaustion, but rather it should be initiated when carbon concentration is close to its half velocity coefficient (Ks) i.e. around 10 g/L. This report demonstrated that the concentration of carbon in the nutrient feed medium used during cultivation is extremely crucial for the biopolymer accumulation. Repeated-batch cultivation is yet another well known technique used for increasing the productivity of the fermentation processes as it eliminates the turnaround time of cleaning, medium preparation and sterilization between the two batch experiments, thereby making the process more productive (Ibrahim and Steinbüchel 2010). In this strategy, a particular volume of the culture broth was removed from the reactor towards the end of cultivation and same volume was immediately replaced with the fresh medium. Recently, Gahlawat and Srivastava (2017) elucidated the use of repeated batch cultivation for the PHB production by *Azohydromonas australica*. During cultivation, a high PHB concentration of 20.55 g/L and PHB productivity of 0.29 g/L h were obtained. This approach showed approximately threefold and twofold increase in PHB productivity and concentrations, respectively as compared to batch culture.

Table 4.3 PHAs production by high cell density fermentation strategies

Fermentation	Microorganism	Type of PHA	Xt (g/L)	PHAc (%)	Qp (g/L h)	References
Fed-batch cultivation	A. eutrophus NCIMB 11599	P(3HB)	164	73	2.42	Kim et al. (1994)
	A. latus DSM 1123	P(3HB)	32.3	70	0.44	Yu et al. (1999)
	A. latus ATCC 29714	P(3HB)	35.4	51	0.99	Grothe and Chisti (2000)
	C. necator NRRL 14690	P(3HB)	39.2	47	0.46	Khanna and Srivastava (2006)
	C. necator DSM 545	P(3HB)	68.8	37	0.84	Cavalheiro et al. (2009)
	C. necator DSM 545	P(3HB)	40	68	0.45	Ienczak et al. (2011)
	A. latus DSM 1123	P(3HB)	18.8	66	0.52	Penloglou et al. (2012)
	C. necator DSM 7237	P(3HB-co-3HV)	27.9	74.5	0.27	Kachrimanidou et al. (2014)
	C. necator H 16	P(3HB-co-3HV)	112.4	83	2.13	Huschner et al. (2015)
	C. necator DSM 545	P(3HB)	125	76	2.03	Mozumder et al. (2014)
	Halomonas venusta KT832796	P(3HB)	37.9	33.4	n/a	Stanley et al. (2017)
	C. necator DSM 545	P(3HB)	148	76	3.1	Haas et al. (2018)
	Halomonas bluephagenesis TD01	P(3HB-co-4HB)	90	74	1.85	Ye et al. (2018)
	Halomonas bluephagenesis TD01	P(3HB)	90	79	1.26	Ren et al. (2018)
Continuous cultivation	Cupriavidus necator DSM 428	P(3HB)	27.2	77	0.5	Cruz et al. (2019)
	A. latus ATCC 29714	P(3HB-co-3HV)	n/a	58	n/a	Ramsay et al. (1990)
	C. necator WSH3	P(3HB)	50	73	1.25	Du et al. (2001)
	P. oleovorans ATCC 29347	mcl-PHA	n/a	63	1.06	Jung et al. (2001)
	C. necator NRRL 14690	P(3HB)	27.7	20	0.55	Khanna and Srivastava (2008)

(continued)

Table 4.3 (continued)

| Fermentation | | | | | | |
Microorganism	Type of PHA	Xt (g/L)	PHAc (%)	Qp (g/L h)	References
C. necator DSM 545	P(3HB)	81	77	1.85	Atlic et al. (2011)
Halomonas TD01	P(3HB)	20	65	n/a	Tan et al. (2011)
A. latus 1124	P(3HB-co-3HV)	24.6	75.9	2.18	Gahlawat and Srivastava (2014)
C. necator DSM 545	n/m	81	77	2.31	Koller et al. (2017)

Note: Xt - total biomass (g/L); PHAc -PHA content (%); Qp - PHA productivity (g/L.h); n/a - data not available; n/m - not mentioned

The continuous fermentation processes are of great commercial importance due their high productivities, particularly with the cultures having high maximum specific growth rates (Ienczak et al. 2013). One of the earliest study on continuous cultivation for PHB and poly (3HB-co-3HV) production was carried out by Ramsay et al. (1990) using *A. eutrophus* and *A. latus*. Du et al. (2001) also reported PHB synthesis by *C. necator* using two-stage continuous cultivation technique. The first stage was used for the culture biomass growth, and the second stage featured PHB production under nitrogen limitation. A high cell density of 50 g/L and PHB concentrations of 30.5 g/L were obtained at 0.075 h^{-1} dilution rate with a significant high PHB productivity of 1.23 g/L h. In another report, feasibility of five-stage or multi-stage continuous cultivation for production of PHAs by *C. necator* was investigated (Atlic et al. 2011). The first stage featured the bacterial growth and other four reactors demonstrated PHAs production under nitrogen limitation. At steady state residual biomass, PHB productivity and content were 19 g/L, 1.97 g/L h and 77% of CDW respectively (from the last bioreactor). Thus, a multi-stage continuous high cell density culture can surpass fed-batch cultivation systems in terms of polymer productivity and final product concentrations. More in-depth understanding of the fermentation kinetics of a particular product, life-long stability of microbial culture and reliable high-density cultivation systems for the production will help in successful commercialization of PHAs-based polymers for different products (Chang et al. 2014).

For sustainable and inexpensive production of PHA, high productivities and high concentrations should be achieved using waste materials. Till date, neither pure nor mixed cultures are able to achieve high productivities using waste materials as feedstock (Obruca et al. 2015; Valentino et al. 2017). Thus, the challenge is still common in both the cases. The major obstacles that affect productivity while using waste materials are their diluted nature and presence of inhibitory compounds (Ienczak et al. 2013). This is also applicable in case of mixed cultures where an anaerobic fermentation step is involved to convert complex sugars into fatty acids. Similarly, other waste materials such as lignocellulosic biomass also require pre-processing step to release sugars from complex compounds (Obruca et al. 2015). Utilization of these treated effluents as feed solution in fed-batch cultivations will result in significant increases in the reactor media volume, thereby resulting in reduced productivity. An interesting way to avoid this problem and achieve high productivities would be to adopt cell recycling strategy along with fed-batch cultivations (Ienczak et al. 2013). This approach has been used for PHA accumulation using both mixed microbial culture (Chen et al. 2015) and pure strains (Haas et al. 2018). The latter achieved high cell densities of 148 g/L using external microfiltration module to retain cells and recycle back into the reactor during fed-batch cultivation of *C. necator* (Haas et al. 2018). This cell-recycle strategy successfully utilized low carbon content feed solution of agro-industrial residues like whey and stillage to attain high yields of PHAs. Thus, the cell recycle strategy appears to provide a good opportunity to increase the biomass concentrations and PHAs productivity from renewable waste materials. This strategy help in preventing the microbial cells from escaping the system, and at the same time results in reduced culture volume and high product titers.

4.5 Mathematical Modelling Strategies

Mathematical models serve as an important tools in bioprocess engineering. They help in better understanding of the system, and facilitate process optimization by designing various nutrients feeding strategies in minimum time period without any trial and error approach (Penloglou et al. 2012). For achieving high PHA accumulation during cultivation, it is important to maintain appropriate concentrations of key nutrients i.e. excess carbon source and limited availability of one of the essential nutrients such as nitrogen or phosphorous in the culture medium (Gahlawat and Srivastava 2014). Therefore, an in-depth understanding of the process kinetics is very important for the designing of various feeding strategies for maintenance of appropriate concentrations of substrates. The mathematical model could serve as an important tool for maintaining high carbon and limiting nitrogen concentrations in bioreactor for enhanced PHB accumulation. In some cases, the concept of double-nutrient limitation zones involving both carbon and nitrogen sources has also been studied for synthesis of PHAs (Du et al. 2001). In few cases, utilization of high initial substrate concentrations e.g. sucrose and waste glycerol have resulted in microbial growth inhibition, thus affects the biomass formation and/or product accumulation rates (Gahlawat and Srivastava 2017; Gahlawat and Soni 2017).

In literature, various mathematical models have been developed to understand the microbial behaviour and optimize the PHB synthesis using microbial cultures (Xu et al. 2005; Gahlawat and Srivastava 2013). Most of them are used as a powerful tool to study the systemic behaviour of culture (Penloglou et al. 2012; Porras et al. 2019) and design adequate cultivation strategies (Gahlawat and Srivastava 2014, 2017) for enhanced biomass and PHB accumulation. Interestingly, mathematical models have not only been used for PHB production, but also applied to PHAs copolymers (Špoljarić et al. 2013; Gahlawat and Srivastava 2014). A mathematical model for batch growth kinetic and poly (3HB-co-3HV) production by *B. megaterium* strain was developed by Porras et al. (2019). The developed model served as a bioprocess optimization tool for improving copolymer production and can be used for designing various feeding strategies. A mathematical model for the enhanced accumulation of poly (3HB-co-3HV) by *Azohydromonas australica* was also proposed by our research group (Gahlawat and Srivastava 2014). A high cell concentration of 29.66 and 21.59 g/L poly (3HB-co-3HV) was obtained by the investigators in 50 h of cultivation using model-based feeding of nutrients during fed-batch cultivations. Recently, our group designed two dynamic fed-batch cultivation strategies using developed mathematical model, and model predictions were further tested by experimental implementation of these strategies at bioreactor level (Gahlawat and Srivastava 2017). The fed-batch cultivation under limiting nitrogen accumulated significantly high biomass concentration of 39.2 g/L and PHAs concentration of 29.7 g/L with overall polymer content of 75%. Therefore, it can be hypothesized that mathematical modeling of microorganism and its subsequent computer simulation under different cultivation conditions are effective tools in biochemical engineering which can lead to better understanding and optimization of the microbial processes with minimum experiments.

Shang et al. (2007) observed that high glucose and phosphate concentration levels significantly affected and inhibited the PHB accumulation in high cell density cultures of *R. eutropha*. Therefore, an unstructured model was developed for predicting the microbial growth and PHB production during high cell density cultivation under phosphate limitation. The simulation results demonstrated that the optimum glucose concentration for PHB accumulation by *R. eutropha* was around 9 g/L (Shang et al. 2007). In order to avoid the substrate limitation and/or inhibition problem, it is very important to keep it around the optimum value to reduce production costs. Although mathematical modeling approach for PHB production have been adopted by several researchers, but PHB utilization during cultivation was ignored in all the developed models. This was introduced in the model system by Franz et al. (2011), while developing a structured, cybernetic model for PHB production kinetics in continuous cultivation systems. In-depth analysis of the continuous PHB production process revealed that its utilization can decrease the final product titers at low dilution rates.

A metabolic and polymerization modeling strategy was suggested by Penloglou et al. (2012) to optimize the process parameters and control molecular properties of final product. The developed model for process optimization yielded in a significantly high PHB amount of 11.84 g/L in 25 h (up to 95% CDW of PHB) and the production of PHB with different molecular weight grades. A mathematical model has been proposed using batch kinetics data of *A. latus* (now renamed as *Azohydromonas australica*) for high biomass and PHB accumulation (Gahlawat and Srivastava 2013). Later on, the mathematical model was utilized for the design of different feeding strategies during fed-batch for enhancing PHB production by *A. latus*. It was possible to improve PHB production from 6.24 g/L achieved in batch (Gahlawat and Srivastava 2012) to 29.64 g/L in model based fed-batch cultivation (Gahlawat and Srivastava 2017). This resulted in 3.5-fold increase in PHB productivity (0.6 g/L h vs. 0.17 g/L h in batch) in model-based fed-batch strategy. Horvat et al. (2013) developed a multistage model for PHA production by *Cupriavidus necator* using five continuous stirred tank reactors (5-CSTR) in series. The main motive of the research work was to develop a dynamic mathematical model as a predictive tool for process optimization. PHA accumulation was observed to be partially growth-associated under nitrogen limitation, therefore the Luedeking-Piret's model of partial growth-associated product formation was adopted as working model. Various model predictions suggested that the PHB production rates can be improved from 2.13 to 10.0 g/L h, if different dilution rates and feed concentrations are used. Mozumder et al. (2014) designed a three-stage optimal feeding strategy for PHB production by *C. necator* DSM 545 during fed-batch cultivation. A high biomass and PHB concentrations of 164 and 125 g/L, respectively was achieved under nitrogen limitation when the glucose level was controlled at optimum range of 9 g/L. The developed optimal feeding strategy was then experimentally validated for waste glycerol which ultimately yielded PHB concentration of 65.6 g/L and PHB content of 62.7% of CDW. Thus it was established that proposed strategy is robust, economic, and sensitive enough for application on fed-batch cultivation.

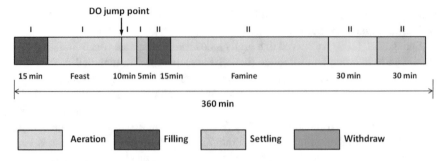

Fig. 4.3 The scheme of one operating cycle for ADD feeding mode. Adopted from Chen et al. (2016) with permission from Elsevier

A simple dynamic metabolic model was developed under feast-famine aerobic dynamic discharge (ADD) mode for the selection of mixed cultures capable of accumulating PHA (Chen et al. 2016). The model included three different types of bacteria on the basis of PHA-accumulation ability, and three model kinetics equation of substrate uptake rate, PHA degradation rate and PHA synthesis rate along with PHA inhibition. The fast PHA accumulation rate achieved in ADD mode was due to the both physical and ecological selective pressure applied on the system by rapid settling and withdrawal after the feeding step. In this way, the mixed microbial culture quickly converted the substrate into intracellular PHA polymer. The scheme of one operating cycle of ADD mode is depicted in Fig. 4.3. More recently, a new technique of specific growth rate and specific production rate estimation was established based upon footprint area analysis of pictures obtained by electron microscopy (Vadija et al. 2016; Koller et al. 2017). This group developed a formal kinetic and structured kinetic model, accompanied by footprint area analysis of binary imaged cells in multistage bioreactor cascade. A continuous five-stage bioreactor cascade mimicked the process features of tubular plug flow reactors, and was developed for high-throughput PHA production by *Cupriavidus necator* at high volumetric and specific PHA productivity of 2.31 and 0.105 g/g h, respectively. Formal kinetic modeling optimized the fermentation process in terms of reactor volumes, dilution rate, and substrate concentration, whereas high structured metabolic model was based on elementary flux analysis depicting the metabolic states of the cell at different prospective scenarios. Footprint area analysis helped in understanding of the morphological changes of the cells and PHA granules under fluctuating environmental conditions (Fig. 4.4). This will facilitate the design of the cascade on a larger scale continuous cultivation, e.g., regarding the optimum number of stages.

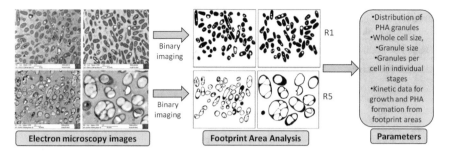

Fig. 4.4 The footprint area analysis of transmission electron microscopy images of *C. necator* DSM 545 culture (*left group of four pictures*) during first stage (R1) and fifth stage (R5) of cultivation in bioreactor cascade. Adopted from Vadija et al. (2016) with permission from Springer

4.6 Sustainable Downstream Recovery Approaches

The extraction of intracellular PHAs from cellular biomass creates a major drawback in the development of a commercially viable fermentation process. This could mainly be due to several reasons such as intracellular nature of product and less solubility in various classical non-toxic solvents. PHAs biopolymers are generally soluble in toxic halogenated solvents such as chloroform which are either expensive or not so easy to handle due to their toxicity (Jacquel et al. 2008). Therefore, simple, economical and effective methods of isolating pure biopolymer from the cells are desperately needed. An ideal extraction protocol must result in high recovery and purity levels at reasonably low cost. Several recovery protocols have been used by different investigators for the extraction of PHAs granules from the cells. Table 4.4 shows the list of different extraction strategies adopted in literature in terms of their merits and demerits. In this section, attempt has been made to summarize the economic and sustainable recovery protocols for PHAs extraction.

The solvent extraction protocol is one of the most extensively adopted methods for PHAs recovery because of its simplicity, recovery efficiency and polymer purity obtained. The solvent extraction protocol is able to remove endotoxins from recovered polymer and cause negligible degradation to PHAs which is extremely useful in medical applications. In the past, the solvent extraction technique was employed by Chemie Linz, an Austrian Company for the recovery of PHB from *A. latus* cells using methylene chloride (Hänggi 1990). However, it has been reported that if the recovered polymer solution contains more than 5% PHB (w/v) then separation of cell residues becomes tricky due to high viscosity of the solution (Jacquel et al. 2008). Therefore, this extraction method requires large quantities of solvents to dissolve the polymer and remove the cell debris. In this case, if solvent recycling is not performed, then recovery process can become expensive (Yang et al. 2011). Moreover, solvents used in the process are toxic and volatile which leads to severe environmental and health issues. Fernández-Dacosta et al. (2015) performed the techno-economic analysis of three downstream processing routes and confirmed that the solvent based recovery process yields the highest cost (1.95 €/kg PHB) and environmental impact.

Table 4.4 Comparison of polyhydroxyalkanoates (PHAs) extraction methods (Jacquel et al. 2008; Kourmentza et al. 2017)

Extraction methods	Merits	Demerits	References
Halogenated solvent	Endotoxin removal, high purity, molecular weight intact	Hazards connected with halogenated solvents, high price, break PHA granules morphology	Jiang et al. (2006), Yang (2011)
Organic solvents	Good recovery; high purity; high molecular weight	Non-recyclable, human toxicity	Aramvash et al. (2015, 2016, 2018)
Non-halogenated solvent	Environment friendly, non-toxic, reusability, recyclability	Low recovery yield at large scale	Fei et al. (2016), Jiang et al. (2018), Gahlawat and Soni (2019)
Surfactant digestion	Extracted polymer retains original molecular weight and native morphology	Low purity, water waste treatment needed	Kim et al. (2003), Mannina et al. (2019)
NaOCl digestion	High purity of PHAs	Degradation of the polymer	Villano et al. (2014)
NaOCl and chloroform	Low polymer degradation, high purity	Large quantity of solvent needed	Ryu et al. (2000)
NaOCl and surfactant	Limited degradation, low operating cost	Large quantity of wastewater	Dong and Sun (2000)
Acid (H_2SO_4) digestion	Defined mechanical strength of PHA	Toxicity issues	López-Abelairas et al. (2015)
Mechanical digestion	No chemicals used, eco-friendly	Time consuming	Arikawa et al. (2017)
Enzymatic digestion	Good recovery, environment friendly	High cost of enzymes, complex process	Lakshman and Shamala (2006), Kachrimanidou et al. (2016)
Biological digestion	No chemical used, low operating cost, molecular weight intact, energy saving	Can contain cell debris and endotoxin	Martínez et al. (2013), Murugan et al. (2016), Ong et al. (2018)
Spontaneous liberation	No extracting chemicals needed	Low recovery (~80% cells secretes PHB granules spontaneously)	Jung et al. (2005)
Dissolved air flotation	No chemicals used	Require several consecutive flotation steps	van Hee et al. (2006)

(continued)

Table 4.4 (continued)

Extraction methods	Merits	Demerits	References
Aqueous two phase system	Safe for separation of sensitive biomaterials, environment friendly, easy scale-up	–	Yeh and Lan (2014), Leong et al. (2018)
Controlled lysis system	Low recovery cost, high purity	Low recovery	Martínez et al. (2011), Israni et al. (2018)
Secretion of PHA granules	Low downstream processing cost	Partial PHA secretion	Rahmann et al. (2013)
Cloud-point extraction	Reusability of phase forming component, high recovery yield	Product quality affected	Leong et al. (2017)

The alkali treatment is more economic process than solvent based recovery with an overall production cost of 1.4 €/kg PHB.

In recent years, researchers are now exploring various non-halogenated solvent such as propylene carbonate, ethanol, butyl acetate and ethyl acetate etc. as an environment friendly alternative to toxic halogenated solvents for the extraction of PHAs. Gahlawat and Soni (2017) reported that 1,2-propylene carbonate could be used as a solvent for the recovery of PHAs from cells of *C. necator*. This group evaluated the effect of temperature and incubation time on PHAs recovery yield, purity and its properties. The highest yield of 90% and purity 93% was obtained under the optimum conditions of purification protocol (Temperature-120 °C, contact time-30 min). 1,2-propylene carbonate exhibits high boiling point of 240 °C which help in prevention of its evaporation into environment at low temperature, and thus can be reused again for various purification cycles (Fiorese et al. 2009). Another study aimed at developing an environmental-friendly and halogen-free approach for the extraction of PHA from genetically engineered *C. necator* using water and ethanol (Mohammadi 2012). On comparison of different incubation conditions, it was observed that the optimized strategy of 1 h incubation at 30 °C resulted in total PHAs purity of 81% and recovery yield of 96%. On the other hand, the chloroform method yielded in PHAs with 100% purity and 95% recovery yield. Therefore, on the basis of these encouraging results with halogen-free approach it was concluded that this method can be used as a potential alternative to halogens method for industrial application at large scale. Riedel et al. (2013) reported a different approach of PHAs recovery using non-halogenated (non-toxic) solvents such as butyl acetate, ethyl acetate and methyl isobutyl ketone. Ethyl acetate exhibited high recovery efficiency and high polymer purity when dry cells were used as starting material. While methyl isobutyl ketone was the most favourable solvent for PHA extraction when wet cells were used. These non-halogenated solvents can also be recycled and reused because of the huge variation in the boiling temperature between extraction solvent and precipitation solvent such as n-heptane. In a recent report, an ecofriendly solvent system of acetone/ethanol/propylene carbonate in ratio of 1:1:1 v/v/v was developed for recovery of PHAs from cells of *C. necator* (Fei et al. 2016). The A/E/P solvent system resulted in maximum PHAs recovery of 85% and PHAs purity of 92% from pre-treated dried biomass. The efficiency of this biorenewable solvent system was further tested on larger scale in a high pressure reactor, and the highest PHAs recovery yield of 65% was obtained at 145 °C. Moreover, the purity and recovery yields were further improved, if hexane was used as precipitant.

Various digestion techniques based on the solubilization of non-PHAs cellular biomass have also been used as an alternative to solvents based recovery protocols (Kunasundari and Sudesh 2011). These can be categorized into three types: chemical (surfactants and sodium hypochlorite) digestion, enzymatic digestion and biological digestion. The importance of chemical digestion method emerges from the fact that surfactants lyse microbial cells without hydrolyzing the biopolymer (Kourmentza et al. 2017). Yang and group (2011) discovered a new approach for the extraction of copolymer, poly (3HB-co-3HV) by linear alkylbenzene sulfonic acid LAS-99 surfactant as a replacement to sodium dodecyl sulfate (SDS) surfactant. LAS-99 is completely biodegradable and environmentally safe compared to SDS. This approach required only 20% of the LAS-99 as opposed to earlier SDS-dependent protocols. Therefore, LAS-99 could prove to be promising in cost-effective and ecofriendly extraction of biopolymer without affecting its properties. Heinrich et al. (2012), proposed a simple and economic recovery approach for large scale extraction of PHB using sodium hypochlorite (NaClO). This strategy yielded a maximum recovery rate of 91.32% with an overall product purity of 96%. Therefore, the evaluation of NaClO treatment has now moved to commercial scales and/or continuous recovery processes. For example, Villano et al. (2014) developed a continuous approach for PHA recovery and it showed high PHAs recovery yield of 100% and purity of 98% using NaClO. This continuous process involved 3 phases: first phase was production using mixed microbial consortia, second was PHAs accumulation, and third was PHAs extraction using NaClO for 24 h. This strategy yielded 1.43 g PHA per litre per day and was stable for four months.

The enzymatic digestion method uses either pure enzymes or crude enzymatic extracts to breakdown the bacterial cell wall for releasing PHAs. This method leads to good recovery of PHAs because of its specificity in catalyzing reaction, but their high cost and complexity of digestion process makes it economically unattractive. Thus, it is essential to focus on the issue of cost reduction to promote the commercial application of enzymatic digestion method. This could be achieved by using immobilized enzymes strategy, integrating enzymes production with PHA biorefinery, and application of genetic engineering approach in enzymes. Likewise, Kachrimanidou et al. (2016) discovered a unique enzyme-based digestion method by utilizing crude extracts obtained after solid-state fermentation of *Aspergillus awamori*. Although this strategy required pre-treatment step of heating and lyophilization, a good P(3HB-co-3HV) recovery yield of 98% and purity of 97% were achieved without adding other chemicals. The bacterial cell lysate, recovered after the removal of polymer, was used as nutrients rich feed solution for PHB production using waste glycerol as main carbon source. Thus, integration of this downstream recovery strategy with sunflower-based biorefinery will assist in reducing the cost of both PHA production medium and extraction strategy.

Biological digestion uses biological agents such as viruses, bacteria and mealworms to release PHA from bacterial cells. The first report on biological digestion method showed utilization of virus particles to lyse the bacterial cells (Schroll et al. 1998). Bacteriophage's lytic cycle capability was utilized to break open bacterial cells to release PHA granules (Madkour et al. 2013). In last few years, researchers have used different biological agents such as predators, rats and mealworms for cell disruption. For e.g. Martínez et al. (2016), reported that *Bdellovibrio bacteriovorus* is an obligate predatory bacteria which can acts as an external cell lytic agent and helps in recovering the intracellular PHA granules from *Pseudomonas putida*. *B. bacteriovorus* exhibited a high recovery yield of 80% and this was attributed to mutated PHA depolymerase gene which otherwise causes unwanted hydrolysis of PHAs. Bacterial PHAs extraction by *B. bacteriovorus* has various benefits such as it helps in avoiding the use of pre-treatment steps and cell harvesting procedure. The researchers have also employed complex organisms digestive system as a non-PHA cell material (NPCM) disruption technique (Murugan et al. 2016). During this technique, mealworms were fed with freeze-dried bacteria and PHAs inclusions were recovered from their feces. This treatment exclusively degraded the NPCM without affecting the molecular weight of polymer. The partially purified PHA inclusions were further purified by water and detergent. The PHAs inclusions showed 89% purity when treated with water alone while purity yield was approximately 100% when reacted with detergent and heat. This method showed no signs of molecular weight reduction and dispersal of the PHA inclusions. The authors elucidated that mealworms feeding on *C. necator* cells exhibited high protein content in comparison to mealworms feeding on oats only. Mealworms feeding on *C. necator* cell can serve as an excellent protein source for aquatic plants and poultry animals. Biological digestion method is an attractive ecofriendly substitute as opposed to other digestion protocols as it does not depend on organic solvents, toxic chemicals and costly equipments for extraction.

Various other methods like Aqueous Two Phase System (ATPS) (Divyashree et al. 2009; Yeh and Lan 2014), liberation of PHA via autolysis (Martínez et al. 2011), secretion of PHA granules (Rahman et al. 2013) and cloud point extraction (Leong et al. 2017) have also been used and demonstrated to have much more success than the other PHA separation protocols. As opposed to conventional solvent extractions methods, ATPS have various advantages and benefits which makes them preferential choice among scientists and industries. The main advantage is that ATPS contain high water content of about 70–90% of w/w, hence they create suitable conditions for the recovery of delicate biopolymers. The polymers that make different layers with ATPS are completely safe and environment friendly as opposed to traditional recovery protocols. Divyashree et al. (2009) reported extraction of PHAs from cell lysate of *Bacillus flexus* using ATPS method. For enzymatic hydrolysis, *B. flexus* cells were mixed with the *Microbispora* sp. culture filtrate having protease and later on subjected to polymer-salt ATPS containing polyethylene glycol and phosphate at pH 8.0 and 28 °C. This method resulted in high molecular weight PHA with an overall purity of 97%. In a recent study, PHA granules were recovered from *C. necator* H16 cells via cloud point extraction (CPE) technique (Leong et al. 2017). It is a type of ATPS strategy that provide the additional advantage to its layer-forming component to be recycled, thereby resulting in cost reduction. The highest PHA recovery yield of around 94% and purity factor of 1.42 was obtained at following conditions of 20% w/w ethylene oxide-propylene oxide (EOPO), 10 mM sodium chloride concentration, and separation temperature of 60 °C with 37.5% w/w of crude feedstock. CPE has been shown to be an excellent procedure for the recovery of PHAs from bacterial crude cell culture.

Furthermore, synthetic biology techniques have also been used to promote the secretion of PHA inclusions from cells into the culture medium. Recently, a self-disruptive and programmed strain of *P. putida* BXHL was designed and constructed from a mcl-PHA producer prototype *P. putida* KT2440 (Martínez et al. 2011). This strain contained controllable self-disruptive system that utilizes endolysin Ejl and holin Ejh proteins for lysis extracted from bacteriophage EJ-1. To further enhance the effectiveness of this cell disruption system, it was assessed in *P. putida tol-pal* mutants that changes in the outer membrane integrity causes lysis hypersensitivity. Based on the findings, it can be established that the programmable cell disruption system of *P. putida* BXHL showed a remarkable method to induce cell lysis during PHAs accumulation, or to produce PHAs harboring cells that were more vulnerable to lysis. This study provided a new outlook to engineered cells promoting PHAs recovery and extraction in a more ecofriendly and economic manner. In another study, the low molecular weight proteins like "Phasins" have been adopted for the release of PHAs from the cells using synthetic biology engineering approach (Rahman et al. 2013). This strategy helped in decreasing the cost associated with the extraction step of PHAs as there was no need to separate the cells for recovery. The Phasins proteins binds to PHB and decreases the size of PHB granules which facilitate release of PHB bound with Phasins from *E. coli* cells via type I secretion using hemolysin (HlyA) signal peptides. However, this study showed that not all the PHB produced is being secreted into the medium, only 36% of the total PHB was released from the cells.

References

Akaraonye E, Keshavarz T, Roy I (2010) Production of polyhydroxyalkanoates: the future green materials of choice. J Chem Technol Biotechnol 85(6):732–743

Alsafadi D, Al-Mashaqbeh O (2017) A one-stage cultivation process for the production of poly-3-(hydroxybutyrate-co-hydroxyvalerate) from olive mill wastewater by *Haloferax mediterranei*. New Biotechnol 34:47–53

Amulya K, Jukuri S, Mohan SV (2015) Sustainable multistage process for enhanced productivity of bioplastics from waste remediation through aerobic dynamic feeding strategy: process integration for up-scaling. Bioresour Technol 188:231–239

Anterrieu S, Quadri L, Geurkink B, Dinkla I, Bengtsson S, Arcos-Hernandez M, Alexandersson T, Morgan-Sagastume F, Karlsson A, Hjort M, Karabegovic L (2014) Integration of biopolymer production with process water treatment at a sugar factory. New Biotechnol 31:308–323

Aramvash A, Gholami-Banadkuki N, Moazzeni-Zavareh F, Hajizadeh-Turchi S (2015) An environmentally friendly and efficient method for extraction of PHB biopolymer with non-halogenated solvents. J Microbiol Biotechnol 25(11):1936–1943

Aramvash A, Gholami-Banadkuki N, Seyedkarimi MS (2016) An efficient method for the application of PHA-poor solvents to extract polyhydroxybutyrate from *Cupriavidus necator*. Biotechnol Prog 32(6):1480–1487

Aramvash A, Moazzeni Zavareh F, Gholami Banadkuki N (2018) Comparison of different solvents for extraction of polyhydroxybutyrate from *Cupriavidus necator*. Eng Life Sci 18(1):20–28

Arikawa H, Sato S, Fujiki T, Matsumoto K (2017) Simple and rapid method for isolation and quantitation of polyhydroxyalkanoate by SDS-sonication treatment. J Biosci Bioeng 124(2):250–254

Aslan AN, Ali MM, Morad NA, Tamunaidu P (2016) Polyhydroxyalkanoates production from waste biomass. IOP Conf Ser Earth Environ Sci 36(1):012040

Atlić A, Koller M, Scherzer D, Kutschera C, Grillo-Fernandes E, Horvat P, Chiellini E, Braunegg G (2011) Continuous production of poly([R]-3-hydroxybutyrate) by *Cupriavidus necator* in a multistage bioreactor cascade. Appl Microbiol Biotechnol 91(2):295–304

Bengtsson S, Karlsson A, Alexandersson T, Quadri L, Hjort M, Johansson P, Morgan-Sagastume F, Anterrieu S, Arcos-Hernandez M, Karabegovic L, Magnusson P (2017) A process for polyhydroxyalkanoate (PHA) production from municipal wastewater treatment with biological carbon and nitrogen removal demonstrated at pilot-scale. New Biotechnol 35:42–53

Berwig KH, Baldasso C, Dettmer A (2016) Production and characterization of poly (3-hydroxybutyrate) generated by *Alcaligenes latus* using lactose and whey after acid protein precipitation process. Bioresour Technol 218:31–37

Bhatia SK, Kim JH, Kim MS, Kim J, Hong JW, Hong YG, Kim HJ, Jeon JM, Kim SH, Ahn J, Lee H (2018) Production of (3-hydroxybutyrate-co-3-hydroxyhexanoate) copolymer from coffee waste oil using engineered *Ralstonia eutropha*. Bioprocess Biosyst Eng 41(2):229–235

Bhattacharyya A, Saha J, Haldar S, Bhowmic A, Mukhopadhyay UK, Mukherjee J (2014) Production of poly-3-(hydroxybutyrate-co-hydroxyvalerate) by *Haloferax mediterranei* using rice-based ethanol stillage with simultaneous recovery and re-use of medium salts. Extremophiles 18(2):463–470

Bugnicourt E, Cinelli P, Lazzeri A, Alvarez V (2014) Polyhydroxyalkanoate (PHA): review of synthesis, characteristics, processing and potential applications in packaging. Express Polym Lett 8:791–808

Cavalheiro JMBT, de Almeida MCMD, Grandfils C, da Fonseca MMR (2009) Poly(3-hydroxybutyrate) production by *Cupriavidus necator* using waste glycerol. Process Biochem 44(5):509–515

Cavalheiro JMBT, Raposo RS, de Almeida MCMD, Teresa CM, Sevrin C, Grandfils C, da Fonseca MMR (2012) Effect of cultivation parameters on the production of poly(3-hydroxybutyrate-co-4-hydroxybutyrate) and poly(3-hydroxybutyrate-4-hydroxybutyrate-3-hydroxyvalerate) by *Cupriavidus necator* using waste glycerol. Biores Technol 111:391–397

Chakravarty P, Mhaisalkar V, Chakrabarti T (2010) Study on poly-hydroxyalkanoate (PHA) production in pilot scale continuous mode wastewater treatment system. Bioresour Technol 101:2896–2899

Chang HN, Jung K, Lee JC, Woo HC (2014) Multi-stage continuous high cell density culture systems: A review. Biotechnol Adv 32(2):514–525

Chanprateep S (2010) Current trends in biodegradable polyhydroxyalkanoates. J Biosci Bioeng 110(6):621–632

Chen GQ, Jiang XR (2018) Engineering microorganisms for improving polyhydroxyalkanoate biosynthesis. Curr Opin Biotechnol 53:20–25

Chen Z, Huang L, Wen Q, Guo Z (2015) Efficient polyhydroxyalkanoate (PHA) accumulation by
    a new continuous feeding mode in three-stage mixed microbial culture (MMC) PHA production
    process. J Biotechnol 209:68–75
Chen Z, Guo Z, Wen Q, Huang L, Bakke R, Du M (2016) Modeling polyhydroxyalkanoate (PHA)
    production in a newly developed aerobic dynamic discharge (ADD) culture enrichment process.
    Chem Eng J 298:36–43
Chen X, Yin J, Ye J, Zhang H, Che X, Ma Y, Li M, Wu LP, Chen GQ (2017) Engineering
    Halomonas bluephagenesis TD01 for non-sterile production of poly (3-hydroxybutyrate-co-4-
    hydroxybutyrate). Bioresour Technol 244:534–541
Ciesielski S, Możejko J, Pisutpaisal N (2015) Plant oils as promising substrates for polyhydrox-
    yalkanoates production. J Clean Prod 106:408–421
Colombo B, Favini F, Scaglia B, Sciarria TP, D'Imporzano G, Pognani M, Alekseeva A, Eisele
    G, Cosentino C, Adani F (2017) Enhanced polyhydroxyalkanoate (PHA) production from the
    organic fraction of municipal solid waste by using mixed microbial culture. Biotechnol Biofuels
    10(1):201
Cruz MV, Gouveia AR, Dionísio M, Freitas F, Reis MA (2019) A process engineering approach to
    improve production of P(3HB) by *Cupriavidus necator* from used cooking oil. Int J Polym Sci
Cui YW, Zhang HY, Lu PF, Peng YZ (2016) Effects of carbon sources on the enrichment of halophilic
    polyhydroxyalkanoate-storing mixed microbial culture in an aerobic dynamic feeding process.
    Sci Rep 6:30766
de Paula FC, de Paula CB, Gomez JGC, Steinbüchel A, Contiero J (2017) Poly (3-hydroxybutyrate-
    co-3-hydroxyvalerate) production from biodiesel by-product and propionic acid by mutant strains
    of *Pandoraea* sp. Biotechnol Prog 33(4):1077–1084
Dhangdhariya JH, Dubey S, Trivedi HB, Pancha I, Bhatt JK, Dave BP, Mishra S (2015) Polyhydrox-
    yalkanoate from marine Bacillus megaterium using CSMCRI's Dry Sea Mix as a novel growth
    medium. Int J Biol Macromol 76:254–261
Dircks K, Beun JJ, Van Loosdrecht M, Heijnen JJ, Henze M (2001) Glycogen metabolism in aerobic
    mixed cultures. Biotechnol Bioeng 73:85–94
Divyashree M, Shamala T, Rastogi N (2009) Isolation of polyhydroxyalkanoate from hydrolyzed
    cells of *Bacillus flexus* using aqueous two-phase system containing polyethylene glycol and
    phosphate. Biotechnol Bioproc Eng 14(4):482–489
Dong Z, Sun X (2000) A new method of recovering polyhydroxyalkanoate from *Azotobacter
    chroococcum*. Chin Sci Bull 45(3):252–256
Du G, Chen J, Yu J, Lun S (2001) Continuous production of poly-3-hydroxybutyrate by *Ralstonia
    eutropha* in a two-stage culture system. J Biotechnol 88(1):59–65
Du C, Sabirova J, Soetaert W, Ki CLS (2012) Polyhydroxyalkanoates production from low-cost
    sustainable raw materials. Curr Chem Biol 6:14–25
Duque AF, Oliveira CS, Carmo IT, Gouveia AR, Pardelha F, Ramos AM, Reis MA (2014) Response
    of a three-stage process for PHA production by mixed microbial cultures to feedstock shift: impact
    on polymer composition. New Biotechnol 31:276–288
Fei T, Cazeneuve S, Wen Z, Wu L, Wang T (2016) Effective recovery of poly-β-hydroxybutyrate
    (PHB) biopolymer from Cupriavidus necator using a novel and environmentally friendly solvent
    system. Biotechnol Prog 38:678–685
Fernández-Dacosta C, Posada JA, Kleerebezem R, Cuellar MC, Ramirez A (2015) Micro-
    bial community-based polyhydroxyalkanoates (PHAs) production from wastewater: techno-
    economic analysis and ex-ante environmental assessment. Bioresour Technol 185:368–377
Fiorese ML, Freitas F, Pais J, Ramos AM, de Aragão GM, Reis MA (2009) Recovery of polyhydrox-
    ybutyrate (PHB) from *Cupriavidus necator* biomass by solvent extraction with 1, 2-propylene
    carbonate. Eng Life Sci 9(6):454–461
Fradinho JC, Domingos JMB, Carvalho G, Oehmen A, Reis MAM (2013) Polyhydroxyalkanoates
    production by a mixed photosynthetic consortium of bacteria and algae. Bioresour Technol
    132:146–153

Franz A, Song HS, Ramkrishna D, Kienle A (2011) Experimental and theoretical analysis of poly (β-hydroxybutyrate) formation and consumption in *Ralstonia eutropha*. Biochem Eng J 55(1):49–58

Gahlawat G, Kumar Soni S (2019) Study on sustainable recovery and extraction of Polyhydroxyalkanoates (PHAs) produced by *Cupriavidus necator* using waste glycerol for medical applications. Chem Biochem Eng Q 33(1):99–110

Gahlawat G, Soni SK (2017) Valorization of waste glycerol for the production of poly (3-hydroxybutyrate) and poly (3-hydroxybutyrate-co-3-hydroxyvalerate) copolymer by *Cupriavidus necator* and extraction in a sustainable manner. Bioresour Technol 243:492–501

Gahlawat G, Srivastava A (2012) Estimation of fundamental kinetic parameters of Polyhydroxybutyrate fermentation process of *Azohydromonas australica* using statistical approach of media optimization. Appl Biochem Biotechnol 168(5):1051–1064

Gahlawat G, Srivastava A (2013) Development of a mathematical model for the growth associated Polyhydroxybutyrate fermentation by *Azohydromonas australica* and its use for the design of fed-batch cultivation strategies. Bioresour Technol 137:98–105

Gahlawat G, Srivastava A (2014) Microbial production of PHB and its copolymers (Ph.D. thesis). Indian Institute of Technology Delhi, India

Gahlawat G, Srivastava A (2017) Model-based nutrient feeding strategies for the increased production of Polyhydroxybutyrate (PHB) by *Alcaligenes latus*. Appl Biochem Biotechnol 183:530–542

García IL, López JA, Dorado MP, Kopsahelis N, Alexandri M, Papanikolaou S, Villar MA, Koutinas AA (2013) Evaluation of by-products from the biodiesel industry as fermentation feedstock for poly (3-hydroxybutyrate-co-3-hydroxyvalerate) production by *Cupriavidus necator*. Biores Technol 130:16–22

Grothe E, Chisti Y (2000) Poly(β-hydroxybutyric acid) thermoplastic production by *Alcaligenes latus*: behavior of fed-batch cultures. Bioprocess Eng 22(5):441–449

Haas R, Jin B, Zepf FT (2008) Production of poly (3-hydroxybutyrate) from waste potato starch. Biosc Biotechnol Biochem 72(1):253–256

Haas C, El-Najjar T, Virgolini N, Smerilli M, Neureiter M (2018) High cell-density production of poly (3-hydroxybutyrate) in a membrane bioreactor. New Biotechnol 37:117–122

Hänggi U (1990) Pilot scale production of PHB with *Alcaligenes latus*. In: Dawes E (ed) Novel biodegradable microbial polymers. Springer, Netherlands, pp 65–70

Heinrich D, Madkour MH, Al-Ghamdi MA, Shabbaj II, Steinbüchel A (2012) Large scale extraction of poly (3-hydroxybutyrate) from *Ralstonia eutropha* H16 using sodium hypochlorite. AMB Express 2(1):59

Hermann-Krauss C, Koller M, Muhr A, Fasl H, Stelzer F, Braunegg G (2013) Archaeal production of polyhydroxyalkanoate (PHA) co-and terpolyesters from biodiesel industry-derived by-products. Archaea

Herrema M, Kimmel K (2012) Method for producing polyhydroxyalkanoic acid. US Patent 8,263,373

Horvat P, Špoljarić IV, Lopar M, Atlić A, Koller M, Braunegg G (2013) Mathematical modelling and process optimization of a continuous 5-stage bioreactor cascade for production of poly [-(R)-3-hydroxybutyrate] by *Cupriavidus necator*. Bioproc Biosyst Eng 36(9):1235–1250

Huschner F, Grousseau E, Brigham CJ, Plassmeier J, Popovic M, Rha C, Sinskey AJ (2015) Development of a feeding strategy for high cell and PHA density fed-batch fermentation of *Ralstonia eutropha* H16 from organic acids and their salts. Process Biochem 50:165–172

Ibrahim MHA, Steinbüchel A (2010) High-cell-density cyclic fed-batch fermentation of a Poly(3-Hydroxybutyrate)-accumulating thermophile, *Chelatococcus* sp. Strain MW10. Appl Environ Microbiol 76(23):7890–7895

Ienczak J, Quines L, Melo AD, Brandellero M, Mendes C, Schmidell W, Aragão G (2011) High cell density strategy for poly (3-hydroxybutyrate) production by *Cupriavidus necator*. Braz J Chem Eng 28(4):585–596

Ienczak JL, Schmidell W, De Aragão GMF (2013) High-cell-density culture strategies for polyhydroxyalkanoate production: a review. J Ind Microbiol Biotechnol 40:275–286

Israni N, Thapa S, Shivakumar S (2018) Biolytic extraction of poly (3-hydroxybutyrate) from *Bacillus megaterium* Ti3 using the lytic enzyme of *Streptomyces albus* Tia1. J Genet Eng Biotechnol 16(2):265–271

Jacquel N, Lo C-W, Wei Y-H, Wu H-S, Wang SS (2008) Isolation and purification of bacterial poly(3-hydroxyalkanoates) Biochem Eng J 39(1):15–27

Jia Q, Xiong H, Wang H, Shi H, Sheng X, Sun R, Chen G (2014) Production of polyhydroxyalkanoates (PHA) by bacterial consortium from excess sludge fermentation liquid at laboratory and pilot scales. Bioresour Technol 171:159–167

Jiang X, Ramsay JA, Ramsay BA (2006) Acetone extraction of mcl-PHA from *Pseudomonas putida* KT2440. J Microbiol Methods 67(2):212–219

Jiang XR, Wang H, Shen R, Chen GQ (2015) Engineering the bacterial shapes for enhanced inclusion bodies accumulation. Metab Eng 29:227–237

Jiang XR, Yao ZH, Chen GQ (2017) Controlling cell volume for efficient PHB production by *Halomonas*. Metab Eng 44:30–37

Jiang G, Johnston B, Townrow D, Radecka I, Koller M, Chaber P, Adamus G, Kowalczuk M (2018) Biomass extraction using non-chlorinated solvents for biocompatibility improvement of polyhydroxyalkanoates. Polymers 10(7):731

Johnson K, Jiang Y, Kleerebezem R, Muyzer G, van Loosdrecht MC (2009) Enrichment of a mixed bacterial culture with a high polyhydroxyalkanoate storage capacity. Biomacromol 10:670–676

Jung K, Hazenberg W, Prieto M, Witholt B (2001) Two-stage continuous process development for the production of medium-chain-length poly(3-hydroxyalkanoates). Biotechnol Bioeng 72:19–24

Jung IL, Phyo KH, Kim KC, Park HK, Kim IG (2005) Spontaneous liberation of intracellular polyhydroxybutyrate granules in *Escherichia coli*. Res Microbiol 156(8):865–873

Kachrimanidou V, Kopsahelis N, Papanikolaou S, Kookos IK, De Bruyn M, Clark JH, Koutinas AA (2014) Sunflower-based biorefinery: Poly(3-hydroxybutyrate) and poly(3-hydroxybutyrate-co-3-hydroxyvalerate) production from crude glycerol, sunflower meal and levulinic acid. Bioresour Technol 172:121–130

Kachrimanidou V, Kopsahelis N, Vlysidis A, Papanikolaou S, Kookos IK, Martínez BM, Rondán MCE, Koutinas AA (2016) Downstream separation of poly(hydroxyalkanoates) using crude enzyme consortia produced via solid state fermentation integrated in a biorefinery concept. Food Bioprod Process 100:323–334

Khanna S, Srivastava AK (2006) Computer simulated fed-batch cultivation for over production of PHB: a comparison of simultaneous and alternate feeding of carbon and nitrogen. Biochem Eng J 27(3):197–203

Khanna S, Srivastava AK (2008) Continuous production of poly-β-hydroxybutyrate by high-cell-density cultivation of *Wautersia eutropha*. J Chem Technol Biotechnol 83(6):799–805

Kim BS, Lee SC, Lee SY, Chang HN, Chang YK, Woo SI (1994) Production of poly(3-hydroxybutyric acid) by fed-batch culture of Alcaligenes eutrophus with glucose concentration control. Biotechnol Bioeng 43(9):892–898

Kim M, Cho K-S, Ryu HW, Lee EG, Chang YK (2003) Recovery of poly (3-hydroxybutyrate) from high cell density culture of *Ralstonia eutropha* by direct addition of sodium dodecyl sulfate. Biotech Lett 25(1):55–59

Koller M, Maršálek L, de Sousa DMM, Braunegg G (2016) Producing microbial polyhydroxyalkanoate (PHA) biopolyesters in a sustainable manner. New Biotechnol 37:24–38

Koller M, Vadija D, Braunegg G, Atlić A, Horvat P (2017) Formal-and high-structured kinetic process modelling and footprint area analysis of binary imaged cells: tools to understand and optimize multistage-continuous PHA biosynthesis. Euro Biotech J 1(3):1–9

Kourmentza C, Ntaikou I, Lyberatos G, Kornaros M (2015) Polyhydroxyalkanoates from *Pseudomonas* sp. using synthetic and olive mill wastewater under limiting conditions. Int J Biol Macromol 74:202–210

Kourmentza C, Plácido J, Venetsaneas N, Burniol-Figols A, Varrone C, Gavala HN, Reis MA (2017) Recent advances and challenges towards sustainable Polyhydroxyalkanoate (PHA) production. Bioeng 4(2):1–43

Kumar P, Kim BS (2019) *Paracoccus* sp. Strain LL1 as a single cell factory for the conversion of waste cooking oil to polyhydroxyalkanoates and carotenoids. Appl Food Biotechnol 6(1):53–60

Kunasundari B, Sudesh K (2011) Isolation and recovery of microbial polyhydroxyalkanoates. Express Polym Lett 5(7):620–634

Lakshman K, Shamala TR (2006) Extraction of polyhydroxyalkanoate from *Sinorhizobium meliloti* cells using *Microbispora* sp. culture and its enzymes. Enzyme Microbial Technol 39(7):1471–1475

Lam W, Wang Y, Chan PL, Chan SW, Tsang YF, Chua H, Yu PHF (2017) Production of polyhydroxyalkanoates (PHA) using sludge from different wastewater treatment processes and the potential for medical and pharmaceutical applications. Environ Technol 38(13–14): 1779–1791

Leong YK, Show PL, Ooi CW, Ling TC, Lan JCW (2014) Current trends in polyhydroxyalkanoates (PHAs) biosynthesis: Insights from the recombinant *Escherichia coli*. J Biotechnol 180:52–65

Leong YK, Lan JCW, Loh HS, Ling TC, Ooi CW, Show PL (2017) Cloud-point extraction of green-polymers from *Cupriavidus necator* lysate using thermoseparating-based aqueous two-phase extraction. J Biosci Bioeng 123:370–375

Leong Y, Chang CK, Arumugasamy S, Lan J, Loh HS, Muhammad D, Show P (2018) Statistical design of experimental and bootstrap neural network modelling approach for thermo separating aqueous two-phase extraction of polyhydroxyalkanoates. Polymers 10(2):132

Li ZJ, Shi ZY, Jian J, Guo YY, Wu Q, Chen GQ (2010) Production of poly (3-hydroxybutyrate-co-4-hydroxybutyrate) from unrelated carbon sources by metabolically engineered *Escherichia coli*. Metab Eng 12(4):352–359

Li S, Cai L, Wu L, Zeng G, Chen J, Wu Q, Chen GQ (2014) Microbial synthesis of functional homo-, random, and block polyhydroxyalkanoates by β-oxidation deleted *Pseudomonas entomophila*. Biomacromol 15:2310–2319

Li T, Ye J, Shen R, Zong Y, Zhao X, Lou C, Chen GQ (2016) Semirational approach for ultra-high Poly(3-hydroxybutyrate) accumulation in *Escherichia coli* by combining one-step library construction and high-throughput screening. ACS Synth Biol 5:1308–1317

Liu S, Abrahamson LP, Scott GM (2012) Biorefinery: ensuring biomass as a sustainable renewable source of chemicals, materials, and energy. Biomass Bioenergy 39:1–4

Loo CY, Sudesh K (2007) Polyhydroxyalkanoates: bio-based microbial plastics and their properties. Malays Polym J 2:31–57

López-Abelairas M, García-Torreiro M, Lú-Chau T, Lema JM, Steinbüchel A (2015) Comparison of several methods for the separation of poly (3-hydroxybutyrate) from *Cupriavidus necator* H16 cultures. Biochem Eng J 93:250–259

Madkour MH, Heinrich D, Alghamdi MA, Shabbaj II, Steinbüchel A (2013) PHA recovery from biomass. Biomacromol 14:2963–2972

Mannina G, Presti D, Montiel-Jarillo G, Suárez-Ojeda ME (2019) Bioplastic recovery from wastewater: a new protocol for polyhydroxyalkanoates (PHA) extraction from mixed microbial cultures. Bioresour Technol 282:361–369

Martinez GA, Bertin L, Scoma A, Rebecchi S, Braunegg G, Fava F (2015) Production of polyhydroxyalkanoates from dephenolised and fermented olive mill wastewaters by employing a pure culture of *Cupriavidus necator*. Biochem Eng J 97:92–100

Martínez V, García P, García JL, Prieto MA (2011) Controlled autolysis facilitates the polyhydroxyalkanoate recovery in *Pseudomonas putida* KT2440. Microb Biotechnol 4:533–547

Martínez V, Jurkevitch E, García JL, Prieto MA (2013) Reward for *Bdellovibrio bacteriovorus* for preying on a polyhydroxyalkanoate producer. Environ Microbiol 15:1204–1215

Martínez V, Herencias C, Jurkevitch E, Prieto MA (2016) Engineering a predatory bacterium as a proficient killer agent for intracellular bio-products recovery: the case of the polyhydroxyalkanoates. Sci Rep 6:24381

Martino L, Cruz MV, Scoma A, Freitas F, Bertin L, Scandola M, Reis MA (2014) Recovery of amorphous polyhydroxybutyrate granules from *Cupriavidus necator* cells grown on used cooking oil. Int J Biol Macromol 71:117–123

Mohammadi M, Hassan MA, Phang LY, Ariffin H, Shirai Y, Ando Y (2012) Recovery and purification of intracellular polyhydroxyalkanoates from recombinant *Cupriavidus necator* using water
and ethanol. Biotechnol Lett 34(2):253–259

Moita R, Lemos PC (2012) Biopolymers production from mixed cultures and pyrolysis by-products.
J Biotechnol 157(4):578–583

Moralejo-Gárate H, Kleerebezem R, Mosquera-Corral A, van Loosdrecht MCM (2011) Microbial
community engineering for biopolymer production from glycerol. Appl Microbiol Biotechnol
92:631–639

Moralejo-Gárate H, Kleerebezem R, Mosquera-Corral A, van Loosdrecht MCM (2013) Impact of
oxygen limitation on glycerol-based biopolymer production by bacterial enrichments. Water Res
47(3):1209–1217

Morgan-Sagastume F, Hjort M, Cirne D, Gérardin F, Lacroix S, Gaval G, Karabegovic L, Alexandersson T, Johansson P, Karlsson A, Bengtsson S (2015) Integrated production of polyhydroxyalkanoates (PHAs) with municipal wastewater and sludge treatment at pilot scale. Biores Technol
181:78–89

Mozumder MSI, De Wever H, Volcke EI Garcia-Gonzalez L (2014) A robust fed-batch feeding
strategy independent of the carbon source for optimal polyhydroxybutyrate production. Process
Biochem 49(3):365–373

Murugan P, Han L, Gan CY, Maurer FH, Sudesh K (2016) A new biological recovery approach for
PHA using mealworm, *Tenebrio molitor*. J Biotechnol 239:98–105

Myung J, Flanagan JC, Waymouth RM, Criddle CS (2017) Expanding the range of polyhydroxyalkanoates synthesized by methanotrophic bacteria through the utilization of omega-
hydroxyalkanoate co-substrates. AMB Express 7(118):1–10

Narayanan A, Kumar VS, Ramana KV (2014) Production and characterization of poly
(3-hydroxybutyrate-co-3-hydroxyvalerate) from *Bacillus mycoides* DFC1 using rice husk
hydrolyzate. Waste Biomass Valorization 5(1):109–118

Nikel PI, De Almeida A, Melillo EC, Galvagno MA, Pettinari MJ (2006) New recombinant
Escherichia coli strain tailored for the production of poly(3-hydroxybutyrate) from agroindustrial
by-products. Appl Environ Microbiol 72(6):3949–3954

Ntaikou I, Peroni CV, Kourmentza C, Ilieva VI, Morelli A, Chiellini E, Lyberatos G (2014) Microbial bio-based plastics from olive-mill wastewater: generation and properties of polyhydroxyalkanoates from mixed cultures in a two-stage pilot scale system. J Biotechnol 188:138–147

Obruca S, Marova I, Snajdar O, Mravcova L, Svoboda Z (2010) Production of poly (3-
hydroxybutyrate-co-3-hydroxyvalerate) by *Cupriavidus necator* from waste rapeseed oil using
propanol as a precursor of 3-hydroxyvalerate. Biotechnol Lett 32(12):1925–1932

Obruca S, Marova I, Melusova S, Mravcova L (2011) Production of polyhydroxyalkanoates from
cheese whey employing *Bacillus megaterium* CCM 2037. Ann Microbiol 61(4):947–953

Obruca S, Benesova P, Marsalek L, Marova I (2015) Use of lignocellulosic materials for PHA
production. Chem Biochem Eng Q 29:135–144

Oh YH, Lee SH, Jang YA, Choi JW, Hong KS, Yu JH, Shin J, Song BK, Mastan SG, David Y,
Baylon MG (2015) Development of rice bran treatment process and its use for the synthesis of
polyhydroxyalkanoates from rice bran hydrolysate solution. Bioresour Technol 181:283–290

Oliveira CS, Silva CE, Carvalho G, Reis MA (2016) Strategies for efficiently selecting PHA producing mixed microbial cultures using complex feedstocks: feast and famine regime and uncoupled
carbon and nitrogen availabilities. New Biotechnol 37:69–79

Ong SY, Zainab-L I, Pyary S, Sudesh K (2018) A novel biological recovery approach for PHA
employing selective digestion of bacterial biomass in animals. Appl Microbiol Biotechnol
102(5):2117–2127

Pais J, Serafim LS, Freitas F, Reis MA (2016) Conversion of cheese whey into
poly (3-hydroxybutyrate-co-3-hydroxyvalerate) by *Haloferax mediterranei*. New Biotechnol
33(1):224–230

Penloglou G, Chatzidoukas C, Kiparissides C (2012) Microbial production of polyhydroxybutyrate with tailor-made properties: an integrated modelling approach and experimental validation. Biotechnol Adv 30(1):329–337

Pfeiffer D, Jendrossek D (2012) Localization of poly(3-Hydroxybutyrate) (PHB) granule-associated proteins during PHB granule formation and identification of two new phasins, phap6 and phap7, in *Ralstonia eutropha* H16. J Bacteriol 194:5909–5921

Phithakrotchanakoon C, Champreda V, Aiba SI, Pootanakit K, Tanapongpipat S (2015) Production of polyhydroxyalkanoates from crude glycerol using recombinant *Escherichia coli*. J Polym Environ 23(1):38–44

Porras MA, Ramos FD, Diaz MS, Cubitto MA, Villar MA (2019) Modeling the bioconversion of starch to P (HB-co-HV) optimized by experimental design using *Bacillus megaterium* BBST4 strain. Environ Technol 40(9):1185–1202

Rahman A, Linton E, Hatch AD, Sims RC, Miller CD (2013) Secretion of polyhydroxybutyrate in *Escherichia coli* using a synthetic biological engineering approach. J Biol Eng 7(24):1–9

Ramachandran H, Amirul AA (2013) Yellow-pigmented *Cupriavidus* sp., a novel bacterium capable of utilizing glycerine pitch for the sustainable production of P(3HB-co-4HB). J Chem Technol Biotechnol 88(6):1030–1038

Ramsay BA, Lomaliza K, Chavarie C, Dube B, Bataille P, Ramsay JA (1990) Production of poly-(beta-hydroxybutyric-co-beta-hydroxyvaleric) acids. Appl Environ Microbiol 56(7):2093–2098

Rao U, Sridhar R, Sehgal PK (2010) Biosynthesis and biocompatibility of poly (3-hydroxybutyrate-co-4-hydroxybutyrate) produced by *Cupriavidus necator* from spent palm oil. Biochem Eng J 49(1):13–20

Rathika R, Janaki V, Shanthi K, Kamala-Kannan S (2018) Bioconversion of agro-industrial effluents for polyhydroxyalkanoates production using *Bacillus subtilis* RS1. Int J Environ Sci Technol. https://doi.org/10.1007/s13762-018-2155-3

Reddy C, Ghai R, Kalia VC (2003) Polyhydroxyalkanoates: an overview. Bioresour Technol 87(2):137–146

Reddy MV, Mawatari Y, Yajima Y, Satoh K, Mohan SV, Chang YC (2016) Production of poly-3-hydroxybutyrate (P3HB) and poly (3-hydroxybutyrate-co-3-hydroxyvalerate) P (3HB-co-3HV) from synthetic wastewater using *Hydrogenophaga palleronii*. Bioresour Technol 215:155–162

Ren Y, Ling C, Hajnal I, Wu Q, Chen GQ (2018) Construction of *Halomonas bluephagenesis* capable of high cell density growth for efficient PHA production. Appl Microbiol Biotechnol 102(10):4499–4510

Riedel SL, Brigham CJ, Budde CF, Bader J, Rha C, Stahl U, Sinskey AJ (2013) Recovery of poly (3-hydroxybutyrate-co-3-hydroxyhexanoate) from *Ralstonia eutropha* cultures with non-halogenated solvents. Biotechnol Bioeng 110(2):461–470

Ryu HW, Cho KS, Lee EG, Chang YK (2000) Recovery of Poly (3-hydroxybutyrate) from coagulated *Ralstonia eutropha* using a chemical digestion method. Biotechnol Prog 16(4):676–679

Salakkam A, Webb C (2018) Production of poly (3-hydroxybutyrate) from a complete feedstock derived from biodiesel by-products (crude glycerol and rapeseed meal). Biochem Eng J 137:358–364

Salgaonkar BB, Mani K, Bragança JM (2019) Sustainable bioconversion of cassava waste to poly (3-hydroxybutyrate-co-3-hydroxyvalerate) by *Halogeometricum borinquense* Strain E3. J Polym Environ 27(2):299–308

Schroll G, Resch S, Gruber K, Wanner G, Lubitz W (1998) Heterologous ΦX174 gene E-expression in *Ralstonia eutropha*: E-mediated lysis is not restricted to γ-subclass of proteobacteria. J Biotechnol 66(2–3):211–217

Serafim LS, Lemos PC, Oliveira RF, Reis MAM (2004) Optimization of polyhydroxybutyrate production by mixed cultures submitted to aerobic dynamic feeding conditions. Biotechnol Bioeng 87:145–160

Shang L, Fan DD, Kim MI, Chang HN (2007) Modeling of poly (3-hydroxybutyrate) production by high cell density fed-batch culture of *Ralstonia eutropha*. Biotechnol Bioproc Eng 12(4):417–423

Shasaltaneh MD, Moosavi-Nejad Z, Gharavi S, Fooladi J (2013) Cane molasses as a source of precursors in the bioproduction of tryptophan by *Bacillus subtilis*. Iran J Microbiol 5:285–292

Solaiman DK, Ashby RD, Hotchkiss JAT, Foglia TA (2006) Biosynthesis of medium-chain-length poly (hydroxyalkanoates) from soy molasses. Biotechnol Lett 28(3):157–162

Špoljarić IV, Lopar M, Koller M, Muhr A, Salerno A, Reiterer A, Malli K, Angerer H, Strohmeier K, Schober S, Mittelbach M (2013) Mathematical modeling of poly [(R)-3-hydroxyalkanoate] synthesis by *Cupriavidus necator* DSM 545 on substrates stemming from biodiesel production. Bioresour Technol 133:482–494

Stanley A, Kumar HP, Mutturi S, Vijayendra SN (2017) Fed-batch strategies for production of PHA using a native isolate of *Halomonas venusta* KT832796 Strain. Appl Biochem Biotechnol. https://doi.org/10.1007/s12010-017-2601-6

Tamis J, Lužkov K, Jiang Y, van Loosdrecht MC, Kleerebezem R (2014) Enrichment of *Plasticicumulans acidivorans* at pilot-scale for PHA production on industrial wastewater. J Biotechnol 192:161–169

Tan D, Xue YS, Aibaidula G, Chen GQ (2011) Unsterile and continuous production of polyhydroxybutyrate by *Halomonas* TD01. Biores Technol 102(17):8130–8136

Tao W, Lv L, Chen GQ (2017) Engineering *Halomonas* species TD01 for enhanced polyhydroxyalkanoates synthesis via CRISPRi. Microb Cell Fact 16(1):48

Tripathi L, Wu LP, Chen J, Chen GQ (2012) Synthesis of diblock copolymer poly-3-hydroxybutyrate-block-poly-3-hydroxyhexanoate [PHB-b-PHHx] by a β-oxidation weakened *Pseudomonas putida* KT2442. Microb Cell Fact 11(44):1–11

Tripathi L, Wu LP, Dechuan M, Chen J, Wu Q, Chen GQ (2013) *Pseudomonas putida* KT2442 as a platform for the biosynthesis of polyhydroxyalkanoates with adjustable monomer contents and compositions. Bioresour Technol 142:225–231

Vadija D, Koller M, Novak M, Braunegg G, Horvat P (2016) Footprint area analysis of binary imaged *Cupriavidus necator* cells to study PHB production at balanced, transient, and limited growth conditions in a cascade process. Appl Microbiol Biotechnol 100(23):10065–10080

Valentino F, Morgan-Sagastume F, Campanari S, Villano M, Werker A, Majone M (2017) Carbon recovery from wastewater through bioconversion into biodegradable polymers. New Biotechnol 37:9–23

van Hee P, Elumbaring AC, van der Lans RG, Van der Wielen LA (2006) Selective recovery of polyhydroxyalkanoate inclusion bodies from fermentation broth by dissolved-air flotation. J Colloid Interface Sci 297(2):595–606

Van Loosdrecht MCM, Pot MA, Heijnen JJ (1997) Importance of bacterial storage polymers in bioprocesses. Water Sci Technol 35:41–47

Verlinden RA, Hill DJ, Kenward MA, Williams CD, Piotrowska-Seget Z, Radecka IK (2011) Production of polyhydroxyalkanoates from waste frying oil by *Cupriavidus necator*. AMB Express 1(1):1–8

Villano M, Valentino F, Barbetta A, Martino L, Scandola M, Majone M (2014) Polyhydroxyalkanoates production with mixed microbial cultures: from culture selection to polymer recovery in a high-rate continuous process. New Biotechnol 31:289–296

Volova TG, Kiselev EG, Shishatskaya EI, Zhila NO, Boyandin AN, Syrvacheva DA, Vinogradova ON, Kalacheva GS, Vasiliev AD, Peterson IV (2013) Cell growth and accumulation of polyhydroxyalkanoates from CO_2 and H_2 of a hydrogen-oxidizing bacterium, *Cupriavidus eutrophus* B-10646. Biores Technol 146:215–222

Wang Y, Yin J, Chen GQ (2014) Polyhydroxyalkanoates, challenges and opportunities. Curr Opin Biotechnol 30:59–65

Wang Y, Chung A, Chen GQ (2017) Synthesis of medium-chain-length polyhydroxyalkanoate homopolymers, random copolymers, and block copolymers by an engineered strain of *Pseudomonas entomophila*. Adv Healthc Mater 6(7):1601017

Wei XX, Shi ZY, Yuan MQ, Chen GQ (2009) Effect of anaerobic promoters on the microaerobic production of polyhydroxybutyrate (PHB) in recombinant *Escherichia coli*. Appl Microbiol Biotechnol 82(4):703–712

Xu J, Guo B, Zhang Z, Wu Q, Zhou Q, Chen J, Chen G, Li G (2005) A mathematical model for regulating monomer composition of the microbially synthesized polyhydroxyalkanoate copolymers. Biotechnol Bioeng 90(7):821–829

Yang YH, Brigham C, Willis L, Rha C, Sinskey A (2011) Improved detergent-based recovery of polyhydroxyalkanoates (PHAs). Biotechnol Lett 33(5):937–942

Bhatia SK, Gurav R, Choi TR, Jung HR, Yang SY, Song HS, Jeon JM, Kim JS, Lee YK, Yang, YH (2019) Poly (3-hydroxybutyrate-co-3-hydroxyhexanoate) production from engineered *Ralstonia eutropha* using synthetic and anaerobically digested food waste derived volatile fatty acids. Int J Biol Macromol

Ye J, Huang W, Wang D, Chen F, Yin J, Li T, Zhang H, Chen GQ (2018) Pilot scale-up of poly (3-hydroxybutyrate-co-4-hydroxybutyrate) production by halomonas bluephagenesis via cell growth adapted optimization process. Biotechnol J 13(5):1800074

Yeh CY, Lan JCW (2014) Direct recovery of polyhydroxyalkanoates synthase from recombinant *Escherichia coli* feedstock by using aqueous two-phase systems. J Taiwan Inst Chem Eng 45(4):1119–1125

Yu PH, Chua H, Huang AL, Ho KP (1999) Conversion of industrial food wastes by *Alcaligenes latus* into polyhydroxyalkanoates. Appl Biochem Biotechnol 78(1–3):445–454

Yue H, Ling C, Yang T, Chen X, Chen Y, Deng H, Wu Q, Chen J, Chen GQ (2014) A seawater-based open and continuous process for polyhydroxyalkanoates production by recombinant *Halomonas campaniensis* LS21 grown in mixed substrates. Biotechnol Biofuels 7(108):1–12

Zafar M, Kumar S, Kumar S, Dhiman AK (2012a) Optimization of polyhydroxybutyrate (PHB) production by *Azohydromonas lata* MTCC 2311 by using genetic algorithm based on artificial neural network and response surface methodology. Biocatal Agr Biotechnol 1(1):70–79

Zafar M, Kumar S, Kumar S, Dhiman AK (2012b) Artificial intelligence based modeling and optimization of poly (3-hydroxybutyrate-co-3-hydroxyvalerate) production process by using *Azohydromonas lata* MTCC 2311 from cane molasses supplemented with volatile fatty acids: A genetic algorithm paradigm. Bioresour Technol 104:631–641

Zafar M, Kumar S, Kumar S, Dhiman AK (2012c) Modeling and optimization of poly (3hydroxybutyrate-co-3hydroxyvalerate) production from cane molasses by *Azohydromonas lata* MTCC 2311 in a stirred-tank reactor: effect of agitation and aeration regimes. J Ind Microbiol Biotechnol 39(7):987–1001

Zúñiga C, Morales M, Le Borgne S, Revah S (2011) Production of poly-β-hydroxybutyrate (PHB) by *Methylobacterium organophilum* isolated from a methanotrophic consortium in a two-phase partition bioreactor. J Hazard Mater 190:876–882

Chapter 5
Summary and Future Perspectives

Abstract This chapter mainly focuses on the future perspectives and the recent developments in strategies for sustainable production of PHA.

Keywords Market penetration · Commercialization · Mixed cultures · Process optimization · High productivity · Low recovery cost

Although there has been a continuous growth on PHA production over the last few years, large-scale applications of PHA have been hindered by the high costs and the lack of process optimization strategies for the high-speed production of PHA. However, present investigations and future developments in this area might well change the market and establish PHA as a mass product similar to other synthetic polymers. Although PHA polymers have been marketed with different trade-names, these attempts have not been successful in long run. Therefore, researchers are still working to develop new bioprocess technologies to further broaden the application of PHA-based commodities, therefore making it cost-effective.

Pure cultures are constantly been exploited for their potential to valorize waste by-products from various industries as renewable and inexpensive feedstock. The capability of some bacteria to grow on lignocellulosic biomass for PHA synthesis is a big advantage in bioprocess optimization. PHA synthesis can also be helpful in the bioremediation of oil contaminated sites, as microorganisms have the capability to degrade toxic chemicals and reduce environmental pollution caused by them. More-over, halophiles are recently been explored as they provide various advantages, such as growth on high-salt containing ocean water and possibility of open, unsterile and continuous fermentation. These characteristics will significantly help in reduction of PHA production cost and minimization of the fresh water requirement. Along with this, synthetic biology techniques are going to help in the improvement of polymer productivity, in simplification of downstream recovery processes and in regulating PHA composition for tailor-made synthesis. Development of genetically modified bacterial strains, that can not only utilize renewable carbon feedstock and at the same time yield higher PHA content and productivity, is required at this moment for sustainable biopolymer production. Recently, research efforts have also been focussed on the PHA production using mixed cultures. Mixed microbial culture based cultivation approach offers various advantages such as reduction in the PHA

G. Gahlawat, *Polyhydroxyalkanoates Biopolymers*,
Biobased Polymers, https://doi.org/10.1007/978-3-030-33897-8_5

production costs and the possibility of utilizing a wide range of waste materials, thus MMC-based plants can be incorporated with industrial wastewater treatment plants. Recent progress in different enrichment techniques for PHA accumulation offers new prospects for cost-effective production of PHA by MMC.

With respect to process optimization, the fact that only few reports are available on continuous cultures in the last five years is not a good indication for commercialization purpose, as these systems are significantly useful for obtaining high PHA productivities. This is in fact important for successful large-scale production, and thus requires further investigation. Multi-stage cultivation in fed-batch or continuous mode, sequential batch fermentation, and continuous cell recycle cultivations are some interesting system that offer considerably high productivities and should be further explored for enhanced PHA production. The development of simple, economic and efficient downstream recovery strategies is essential for the sustainable production of PHAs. Recently, new strategies based on the utilization of environmentally friendly techniques such as aqueous two phase system, spontaneous liberation of PHA, autolysis of cells are constantly being investigated. Utilization of recyclable and reusable non-halogenated (non-toxic) solvents, and production of enzymes as part of a PHA biorefinery can reduce the downstream recovery cost. Thus, an amalgamation of appropriate process design, inexpensive substrates, and economic sustainable recovery strategies are going to facilitate their large-scale production.

Printed in the United States
By Bookmasters